Universitext

Universitext

Editors (North America): J.H. Ewing, F.W. Gehring, and P.R. Halmos

Aksoy/Khamsi: Nonstandard Methods in Fixed Point Theory
Aupetit: A Primer on Spectral Theory
Booss/Bleecker: Topology and Analysis
Carleson/Gamelin: Complex Dynamics
Cecil: Lie Sphere Geometry: With Applications to Submanifolds
Chae: Lebesgue Integration (2nd ed.)
Charlap: Bieberbach Groups and Flat Manifolds
Chern: Complex Manifolds Without Potential Theory
Cohn: A Classical Invitation to Algebraic Numbers and Class Fields
Curtis: Abstract Linear Algebra
Curtis: Matrix Groups
DiBenedetto: Degenerate Parabolic Equations
Dimca: Singularities and Topology of Hypersurfaces
Edwards: A Formal Background to Mathematics I a/b
Edwards: A Formal Background to Mathematics II a/b
Foulds: Graph Theory Applications
Gardiner: A First Course in Group Theory
Gårding/Tambour: Algebra for Computer Science
Goldblatt: Orthogonality and Spacetime Geometry
Hahn: Quadratic Algebras, Clifford Algebras, and Arithmetic Witt Groups
Holmgren: A First Course in Discrete Dynamical Systems
Howe/Tan: Non-Abelian Harmonic Analysis: Applications of *SL*(2, R)
Humi/Miller: Second Course in Ordinary Differential Equations
Hurwitz/Kritikos: Lectures on Number Theory
Jennings: Modern Geometry with Applications
Jones/Morris/Pearson: Abstract Algebra and Famous Impossibilities
Kelly/Matthews: The Non-Euclidean Hyperbolic Plane
Kostrikin: Introduction to Algebra
Luecking/Rubel: Complex Analysis: A Functional Analysis Approach
MacLane/Moerdijk: Sheaves in Geometry and Logic
Marcus: Number Fields
McCarthy: Introduction to Arithmetical Functions
Meyer: Essential Mathematics for Applied Fields
Mines/Richman/Ruitenburg: A Course in Constructive Algebra
Moise: Introductory Problems Course in Analysis and Topology
Morris: Introduction to Game Theory
Porter/Woods: Extensions and Absolutes of Hausdorff Spaces
Ramsay/Richtmyer: Introduction to Hyperbolic Geometry
Reisel: Elementary Theory of Metric Spaces
Rickart: Natural Function Algebras
Rotman: Galois Theory
Sagan: Space-Filling Curves
Samelson: Notes on Lie Algebras
Schiff: Normal Families
Shapiro: Composition Operators and Classical Function Theory
Smith: Power Series From a Computational Point of View
Smoryński: Self-Reference and Modal Logic
Stillwell: Geometry of Surfaces
Stroock: An Introduction to the Theory of Large Deviations
Sunder: An Invitation to von Neumann Algebras
Tondeur: Foliations on Riemannian Manifolds

Shiing-shen Chern

Complex Manifolds Without Potential Theory

(with an appendix on the geometry of characteristic classes)

Second Edition

Revised Printing

Springer-Verlag
New York Berlin Heidelberg London Paris
Tokyo Hong Kong Barcelona Budapest

Shiing-shen Chern
Department of Mathematics
University of California
Berkeley, CA 94720
USA

Mathematics Subject Classifications (1991): 32C10, 53B35

Library of Congress Cataloging-in-Publication Data
Chern, Shiing-shen, 1911-
Complex manifolds without potential theory : with an appendix on the geometry of characteristic classes / Shiing-shen Chern. -- 2nd ed.
p. cm. -- (Universitext)
"Revised printing."
Includes bibliographical references and index.
ISBN 0-387-90422-0 (acid free)
1. Complex manifolds. 2. Geometry, Differential. I. Title.
QA331.C45 1995
515'.9223--dc20 94-41846

Printed on acid-free paper.

The First Edition was published in 1968 by van Nostrand.

Production managed by Jim Harbison; manufacturing supervised by Genieve Shaw.
Printed and bound by Braun-Brumfield, Ann Arbor, MI.
Printed in the United States of America.

9 8 7 6 5 4 3 2 (Revised printing)

ISBN 0-387-90422-0 Springer-Verlag New York Berlin Heidelberg
ISBN 3-540-90422-0 Springer-Verlag Berlin Heidelberg New York

Preface

The main text on complex manifolds was the notes from a course with the same title at the University of California Los Angeles in the fall of 1966. It was written up after each lecture; only minor changes have been made. To the Department of Mathematics at UCLA, and Lowell Paige in particular, I wish to express here my belated thanks. The Appendix is an expanded version of a series of lectures given at a summer seminar of the Canadian Mathematical Congress that took place in Halifax, Nova Scotia in 1971. I wish to thank Professor J.R. Vanstone for his hospitality and kindness. Noteworthy is the treatment of the secondary characteristic classes, which is different from the one given in [19] in the Bibliography to the Appendix.

Needless to say, my deepest gratitude is to the University of California at Berkeley and the National Science Foundation for their continuous support of my research.

Contents

§1. Introduction and Examples

A complex manifold is a paracompact Hausdorff space which has a covering by neighborhoods each homeomorphic to an open set in the m-dimensional complex number space such that where two neighborhoods overlap the local coordinates transform by a complex analytic transformation. That is, if $z^1,\ldots,z^m$ are local coordinates in one such neighborhood and if $w^1,\ldots,w^m$ are local coordinates in another neighborhood, then where they are both defined, we have $w^i = w^i(z^1,\ldots,z^m)$, where each w^i is a holomorphic (or analytic) function of the z's and the functional determinant $\partial(w^1,\ldots,w^m)/\partial(z^1,\ldots,z^m) \neq 0$.

We will give some examples of complex manifolds:

Example 1. The complex number space C_m whose points are the ordered m-tuples of complex numbers $(z^1,\ldots,z^m)$. C_1 is called the Gaussian plane.

Example 2. The complex projective space P_m. To define it, take $C_{m+1} - 0$, where 0 is the point $(0,\ldots,0)$, and identify those points $(z^0,z^1,\ldots,z^m)$ which differ from each other by a factor. The resulting quotient space is P_m. It can be covered by $m+1$ open sets U_i defined respectively by $z^i \neq 0$, $0 \leqq i \leqq m$. In U_i we have the local coordinates ${}_i\zeta^k = z^k/z^i$, $0 \leqq k \leqq m$, $k \neq i$. The transition of local coordinates in $U_i \cap U_j$ is given by ${}_j\zeta^h = {}_i\zeta^h / {}_i\zeta^j$, $0 \leqq h \leqq m$, $h \neq j$, which are holomorphic functions. In particular, P_1 is the Riemann sphere.

By assigning to a point of $C_{m+1} - 0$ the point it defines in the quotient space, we get a natural projection $\psi: C_{m+1} - 0 \to P_m$, for which the inverse image of each point is $C^* = C_1 - 0$. This relationship is the first example of the important notion of a holomorphic line bundle and it is justified to enter into some detail. In fact, in $\psi^{-1}(U_i)$ we can use instead of the coordinates $(z^0,\ldots,z^m)$ the coordinates ${}_i\zeta^h = z^h/z^i$, $0 \leqq h \leqq m$, $h \neq i$, and z^i. This has the advantage of expressing clearly the fact that $\psi^{-1}(U_i)$ is a product $U_i \times C^*$, z^i being the fiber coordinate (relative to U_i). In $\psi^{-1}(U_i \cap U_j)$, the fiber coordinates z^i and z^j, relative to U_i and U_j respectively, are related by

$$z^i = z^j\,{}_j\zeta^i = z^j/{}_i\zeta^j .$$

Thus the change of fiber coordinates is expressed by the multiplication of a non-zero holomorphic function. The general notion of a holomorphic line bundle, which generalizes this example, plays a central rôle in complex manifolds.

To a point $p \in P_m$ the coordinates of a point of $\psi^{-1}(p)$ are called its homogeneous coordinates. They can be normalized so that

$$\Sigma\, z^k \bar{z}^k = 1 . \tag{1.1}$$

Equation (1.1) defines a sphere S^{2m+1} of real dimension $2m+1$. The restriction of ψ gives the mapping $\psi\colon S^{2m+1} \to P_m$, under which the inverse image of each point is a circle. This is called the *Hopf fibering* of S^{2m+1}.

Further examples are obtained from submanifolds of P_m and quotient manifolds of C_m.

Example 3. Non-singular submanifolds of P_m, in particular, the non-singular hyperquadric

$$(z^0)^2 + \ldots + (z^m)^2 = 0 . \tag{1.2}$$

By a *theorem of Chow,* every compact submanifold imbedded in P_m is an algebraic variety, i.e., it is the locus defined by a finite number of homogeneous polynomial equations [5, p. 170].

It will not be significant to consider compact submanifolds of C_m, because of the theorem:

(A) A connected compact submanifold of C_m is a point.

The proof makes use of the lemma: Let f be a holomorphic function on a complex manifold M. Suppose $p_0 \in M$ is a point such that $|f(p)| \leqq |f(p_0)|$ for all p in a neighborhood of p_0. Then $f(p) = f(p_0)$ in a neighborhood of p_0.

For one variable this follows from the maximum modulus principle. The case of m variables follows from the consideration of the lines through p_0 and the application of the one variable case to these lines.

Now let M be a connected compact submanifold of C_m. Each coordinate of C_m is a holomorphic function on M. By the lemma, it must be a constant on every connected component of M. Since M is connected, M is a point.

However, various significant examples arise from the quotient manifolds of C_m:

<u>Example 4</u>. Let Γ be the discontinous group generated by $2m$ translations of C_m, which are linearly independent over the reals. Then C_m/Γ is called the *complex torus*. If a complex torus can be imbedded as a non-singular submanifold of a projective space of sufficiently high dimension, it is called an *abelian variety*.

Let Δ be the discontinuous group generated by $z^k \to 2z^k$, $1 \leqq k \leqq m$. The quotient manifold $(C_m-0)/\Delta$ is called the *Hopf manifold*. It is homeomorphic to $S^1 \times S^{2m-1}$.

Consider C_3 to be the group of all the matrices

$$(1.3) \qquad \begin{pmatrix} 1 & z_1 & z_2 \\ 0 & 1 & z_3 \\ 0 & 0 & 1 \end{pmatrix}$$

Let D be the discrete subgroup consisting of those matrices for which z_1, z_2, z_3 are Gaussian integers (i.e., $z_k = m_k + in_k$, $1 \leqq k \leqq 3$, where m_k, n_k are rational integers). Then C_3/D is called an *Iwasawa manifold*. Its fundamental group is isomorphic to D, and hence is not abelian.

Example 5. An orientable surface is a complex manifold (of dimension one). We suppose the surface to be C^∞ and define on it a positive definite riemannian metric. By the theorem of Korn-Lichtenstein there exist local isothermal parameters x, y so that locally the metric can be written

$$ds^2 = \lambda^2(dx^2 + dy^2), \qquad \lambda > 0 \tag{1.4}$$

or $ds^2 = \lambda^2 dz\, d\bar{z}$, where $z = x + iy$, the orientation of the manifold being defined by $dx \wedge dy = \frac{i}{2}\, dz \wedge d\bar{z}$. If w is another local coordinate we will have

$$ds^2 = \lambda^2 dz\, d\bar{z} = \mu^2 dw\, d\bar{w}$$

because ds^2 is globally defined. It follows that dw is a multiple of dz or $d\bar{z}$. If we assume that the complex coordinates z and w define the same orientation, then dw must be a multiple of dz. This means that w is a holomorphic function of z, and the surface becomes a complex manifold.

A one-dimensional complex manifold is called a *Riemann surface*.

Example 6. (Calabi-Eckmann) Let S and S' be spheres of dimensions 2p+1 and 2q+1 respectively, p, q > 0. By the Hopf fibering in Ex. 2 we have a fibration

$$\pi: S \times S' \to P_p \times P_q$$

with fiber a two (real) dimensional torus. Since both the base space and the fiber are complex manifolds, we would expect that the total space could be given a complex structure. This we will prove to be the case as follows:

Let S be the set of all points $z = (z^0,\dots,z^p)$ such that $\Sigma_{0\leqq k\leqq p}\, z^k\, \bar{z}^k = 1$, and S' be the set of all points $z' = (z'^0,\dots, z'^q)$ such that $\Sigma_{0\leqq j\leqq q}\, z'^j\, \bar{z}'^j = 1$. We define

$$V_{kj} = \{(z,z') \in S \times S' \,|\, z^k\, z'^j \neq 0\}\ ,\quad 0 \leqq k \leqq p,\ 0 \leqq j \leqq q\ .$$

Then the sets V_{kj} form an open covering of $S \times S'$. Let τ be a complex number such that $\operatorname{Im}(\tau) \neq 0$. In V_{kj} we introduce the following complex coordinates

$$\begin{aligned} &{}_k w^h = z^h/z^k\ ,\ {}_j w'^{\ell} = z'^{\ell}/z'^j\ ,\ h \neq k\ ,\ \ell \neq j\ ,\quad 0 \leqq h \leqq p\ , \\ (1.5)\qquad & \qquad\qquad\qquad\qquad\qquad\qquad\qquad\qquad\quad 0 \leqq \ell \leqq q\ , \\ &t_{kj} = \frac{1}{2\pi i}\,(\log z^k + \tau \log z'^j)\ , \end{aligned}$$

where t_{kj} is defined mod 1 and τ. Thus t_{kj} defines a point on the torus $T_{(1,\tau)}$ which is the quotient of C by the translations 1 and τ. In this way we have $p + q + 1$ coordinates in V_{kj} and these define a map $V_{kj} \to C_{p+q} \times T_{(1,\tau)}$. We show that this map is a homeomorphism. It suffices to show that ${}_k w^h$, ${}_j w'^{\ell}$ and t_{kj} determine the z's and z''s uniquely. Now

$$\sum_{h\neq k} {}_k w^h\ {}_k\bar{w}^h = \sum_h \frac{z^h\,\bar{z}^h}{z^k\,\bar{z}^k} - 1 = \frac{1}{|z^k|^2} - 1\ ,$$

so $|z^k|$ is determined. Similarly, $|z'^j|$ is determined. By the second equation of (1.5) we have

$$t_{kj} = \frac{1}{2\pi i}\,(\log|z^k| + \tau \log|z'^j| + i \arg z^k + \tau\, i \arg z'^j)\ ,$$

$$\operatorname{mod}(1,\tau)\ .$$

Hence $\arg z^k$, $\arg z'^j$ are determined mod 2π. The other z's and z''s are then determined by the first equations of (1.5). This proves that our map is a homeomorphism.

In $V_{kj} \cap V_{rs}$ we have

$$ {}_r w^h = {}_k w^h / {}_k w^r \quad , \quad {}_s w'^\ell = {}_j w'^\ell / {}_j w'^s $$

and

$$ t_{rs} = t_{kj} + \frac{1}{2\pi i} (\log {}_k w^r + \tau \log {}_j w'^s) \ , $$

$$ \mathrm{mod}(1,\tau) $$

where we set ${}_k w^k = 1$ and ${}_j w'^j = 1$. Hence we have defined a complex structure on $S \times S'$ with the ${}_k w^h$, ${}_j w'^\ell$, t_{kj} as local coordinates in V_{kj}.

§2. Complex and Hermitian Structures on a Vector Space

Let V be a real vector space of dimension n. V is said to have a complex structure if there exists a linear endomorphism $J: V \to V$, such that $J^2 = -1$, where 1 denotes the identity endomorphism. An eigenvalue of J is a complex number λ such that the equation $Jx = \lambda x$ has a non-zero solution $x \in V$. Applying J to this equation, we get

$$ -x = J^2 x = \lambda J x = \lambda^2 x \ . $$

It follows that $\lambda^2 = -1$ or $\lambda = \pm i$. Since the complex eigenvalues occur in conjugate pairs, it follows that V must be of even dimension $n = 2m$. The following relations are immediately verified:

$$ (J - i)(J + i) = (J + i)(J - i) = 0 \ . \tag{2.1} $$

Let V^* be the dual space of V, i.e., the space of all

real-valued linear functions over V. We denote the pairing of V and V^* by $\langle x, y^* \rangle$, $x \in V$, $y^* \in V^*$, so that this function is R-linear in each of the arguments. Alternatively, we also write $y^*(x) = \langle x, y^* \rangle$. In addition to V^* we consider $V^* \otimes C$, i.e., the space of all complex-valued R-linear functions over V. Then $V^* \otimes C$ is a complex vector space of complex dimension n. An element $f \in V^* \otimes C$ is said to be of type (1,0) (respectively (0,1)) if

(2.2) $$f(Jx) = if(x) \text{ (resp. } f(Jx) = -if(x)), \; x \in V .$$

Let $e^{*\alpha}$, $1 \leqq \alpha \leqq n$, be a basis of V^*. Consider the functions

(2.3) $$\lambda^{\alpha}(x) = \langle (J+i)x, e^{*\alpha} \rangle = \langle Jx, e^{*\alpha} \rangle + i \langle x, e^{*\alpha} \rangle .$$

Since $-i$ is an eigenvalue of J of multiplicity m, exactly m of these functions are linearly independent with respect to C. It can be immediately verified that $\lambda^{\alpha}(x)$ are of type (1,0), and their complex conjugates

(2.3a) $$\bar{\lambda}^{\alpha}(x) = \langle (J-i)x, e^{*\alpha} \rangle$$

are of type (0,1).

Suppose our basis $e^{*\alpha}$ is so chosen that $\lambda^k(x)$, $1 \leqq k \leqq m$, are linearly independent with respect to C. We split them into the real and imaginary parts:

(2.4) $$\lambda^k(x) = \lambda'^k(x) + i\lambda''^k(x) .$$

We wish to show that $\lambda'^k(x)$, $\lambda''^k(x)$, $1 \leqq k \leqq m$, are linearly independent with respect to R. In fact, suppose that

$$\sum_k r_k \lambda'^k(x) + \sum_k s_k \lambda''^k(x) = 0 , \quad x \in V ,$$

where r_k, $s_k \in R$. This relation can be written as

$$\sum_k (r_k - is_k)\lambda^k(x) + \sum_k (r_k + is_k)\bar{\lambda}^k(x) = 0 .$$

Replacing x by Jx and using the fact that $\lambda^k(x)$ and $\bar{\lambda}^k(x)$ are of types (1,0) and (0,1) respectively, we get

$$\sum_k (r_k - is_k)\lambda^k(x) - \sum_k (r_k + is_k)\bar{\lambda}^k(x) = 0 .$$

Adding these two equations, we find

$$\Sigma(r_k - is_k)\lambda^k(x) = 0 .$$

which gives $r_k - is_k = 0$, and hence $r_k = s_k = 0$, since $\lambda^k(x)$ are linearly independent over C.

Using the exterior algebra $\wedge(V^* \otimes C)$, we can express the fact proved above by

$$(2.5) \qquad \left(\frac{i}{2}\right)^m \bigwedge_k \lambda^k\bar{\lambda}^k = \bigwedge_k \lambda'^k\lambda''^k \neq 0 .$$

It follows from (2.5) that λ^k, $\bar{\lambda}^k$ are linearly independent over C and that $V^* \otimes C$ is a direct sum of $V_C \oplus \bar{V}_C$, where V_C (resp. $\bar{V}_C$) is the space of all elements of $V^* \otimes C$ of type (1,0) (resp. (0,1)). Conversely, a direct sum decomposition of $V^* \otimes C$ over C into two subspaces, complex conjugate to each other, defines a complex structure on V, if the subspaces are defined to be consisting of the elements of types (1,0) and (0,1) respectively. This follows from the fact that when x is given the equations in (2.2) determine the values of the elements of $V^* \otimes C$ at Jx, whereby Jx is determined.

As an example, let e_k, e_{m+k} be a dual basis of λ'^k, λ''^k, so that

$$(2.6) \qquad \lambda'^k(e_h) = \lambda''^k(e_{m+h}) = \delta^k_h , \quad 1 \leqq h, k \leqq m ,$$

all other pairings being zero. This can be written

$$(2.6a) \qquad \lambda^k(e_h) = \frac{1}{i}\lambda^k(e_{m+h}) = \delta^k_h .$$

On the other hand, we have

$$\lambda^k(Jx) = i\lambda^k(x) = -\lambda''^k(x) + i\lambda'^k(x) ,$$

from which it follows that

$$(2.7) \qquad \lambda^k(Je_h) = \frac{1}{i}\lambda^k(Je_{m+h}) = i\,\delta^k_h .$$

Comparing (2.6a) and (2.7), we get

$$(2.8) \qquad Je_h = e_{m+h},\ Je_{m+h} = -e_h .$$

The elements λ^k form a basis of V_C over C. Under a change of basis the real-valued 2m-form in (2.5) will be multiplied by a positive factor. Hence the complex structure J in V defines an orientation of V.

If J defines a complex structure in V, $-J$ does too. The two complex structures are said to be *conjugate* to each other. A form of type (1,0) (resp. type (0,1)) in the structure J is a form of type (0,1) (resp. (1,0)) in the structure $-J$ and vice versa.

Suppose V is provided with a complex structure J. An hermitian structure in V is a complex-valued function $H(x,y)$, $x,y \in V$, which satisfies the following conditions:

(1) $H(\lambda_1x_1 + \lambda_2x_2,y) = \lambda_1H(x_1,y) + \lambda_2H(x_2,y)$,

$x_1,x_2,y \in V$, $\lambda_1,\lambda_2 \in R$;

(2) $\overline{H(x,y)} = H(y,x)$

(3) $H(Jx,y) = iH(x,y)$.

In view of (2), (3) is equivalent to the following:

(3') $H(x,Jy) = -iH(x,y)$

We split $H(x,y)$ into its real and imaginary parts:

$$(2.9) \qquad H(x,y) = F(x,y) + iG(x,y) .$$

Then (2) is equivalent to

$$(2.10) \qquad F(x,y) = F(y,x), \; G(x,y) = -G(y,x) ,$$

and (3) is equivalent to

$$(2.11) \qquad F(x,y) = G(Jx,y), \text{ or } G(x,y) = -F(Jx,y) .$$

Thus $H(x,y)$ defines a pair of real-valued bilinear functions, of which one is symmetric and the other anti-symmetric, which are related to each other by (2.11). Either one of these functions, together with J, determines $H(x,y)$.

The hermitian scalar product $H(x,y)$ is called *positive definite* if the corresponding real-valued symmetric bilinear function $F(x,y)$ is positive definite.

If is known that the space $\Lambda^2(V^*)$ of exterior forms of degree two is isomorphic to the space of all antisymmetric bilinear functions over V. The isomorphism is established by the fact that it is a vector space isomorphism and that for $\xi, \eta \in V^*$ the bilinear function corresponding to $\xi \wedge \eta$ is

$$(2.12) \qquad (\xi \wedge \eta)(x,y) = 1/2 \, \{\xi(x)\eta(y) - \xi(y)\eta(x)\}, \; x,y \in V .$$

By means of this isomorphism there is an exterior form $\hat{H}$ of degree two corresponding to the function $-1/2 \, G(x,y)$. $\hat{H}$ is called the *Kähler form* of the hermitian structure.

We wish to express $H(x,y)$ in terms of the basis λ^k of V_C . For this purpose let

$$(2.13) \qquad x = \sum_\alpha x^\alpha e_\alpha , \; y = \sum_\beta y^\beta e_\beta , \; 1 \leqq \alpha,\beta \leqq 2m, \; 1 \leqq k,j \leqq m .$$

Then we have

$$\begin{aligned}
H(x,y) &= H(\sum_k (x^k e_k + x^{m+k} e_{m+k}), y) = H(\sum_k (x^k e_k + x^{m+k} J e_k), y) \\
&= \sum_k (x^k + i x^{m+k}) H(e_k , y) \\
&= \sum_{k,j} (x^k + i x^{m+k})(y^j - i y^{m+j}) H(e_k , e_j) \\
&= \sum_{k,j} \lambda^k(x)\, \bar{\lambda}^j(y) H(e_k , e_j)
\end{aligned}$$

It follows that we can write

$$H = \sum_{k,j} h_{kj} \lambda^k \otimes \bar{\lambda}^j \tag{2.14}$$

where

$$h_{kj} = H(e_k , e_j) = \bar{h}_{jk} \tag{2.15}$$

To find the expression for the Kähler form $\hat{H}$, we derive from (2.9)

$$\begin{aligned}
-1/2(G(x,y)) &= \frac{i}{4} \{H(x,y) - \bar{H}(x,y)\} \\
&= \frac{i}{4} \sum_{k,j} h_{kj} \{\lambda^k(x)\bar{\lambda}^j(y) - \bar{\lambda}^j(x)\lambda^k(y)\}
\end{aligned}$$

By (2.12) it follows that

$$\hat{H} = \frac{i}{2} \sum_{k,j} h_{kj} \lambda^k \wedge \bar{\lambda}^j \tag{2.16}$$

If a real vector space has a complex structure and in addition to it an hermitian structure, the exterior algebra has rich properties. In particular, a complex-valued exterior form, i.e., an element of the exterior algebra $\wedge (V^* \otimes C)$, is said to be of type (p,q) , if it is a sum of terms each of which contains p factors λ^k and q factors $\bar{\lambda}^h$. A form of degree r can be written uniquely as a sum

$$\alpha = \sum_{p+q=r} \alpha_{pq} \, , \quad (p,q) \text{ mutually distinct,} \tag{2.17}$$

where α_{pq} is of type (p,q). The latter will also be denoted by

(2.18) $$\alpha_{pq} = \Pi_{pq} \alpha ,$$

whereby the operators Π_{pq} are defined.

Another operator, which we will denote by L , is defined by

(2.19) $$L\alpha = \hat{H} \wedge \alpha .$$

L is a real operator in the sense that it maps a real-valued form into a real-valued form. This operator L plays an important rôle in Hodge's work on transcendental methods in algebraic geometry.

§3. Almost Complex Manifolds; Integrability Conditions

Let M be a C^∞ manifold of dimension n . To a point $x \in M$ we will denote by T_x and T_x^* the tangent and cotangent spaces respectively. An *almost complex structure* on M is a C^∞ field of endomorphisms $J_x: T_x \to T_x$, such that $J_x^2 = -1_x$, where 1_x denotes the identity endomorphism in T_x .

A manifold which is given an almost complex structure is called *almost complex*. Not all manifolds have this property. In fact, from the discussions in §2 follows the theorem:

(A) An almost complex manifold is even-dimensional and orientable.

Remark. This condition is not sufficient for a manifold to have an almost complex structure. For instance, it was proved by Ehresmann and Hopf that the 4-sphere S^4 cannot be given an almost complex structure [11, p. 217].

Alternatively, an almost complex structure can be defined by the space A of its complex-valued C^∞ forms of type (1,0). If $\bar{A}$ denotes the space consisting of forms which are conjugate complex to those of A , then at every $x \in M$ we have the direct sum decomposition

$$T_x^* \otimes C = A_x \oplus \bar{A}_x , \tag{3.1}$$

where A_x (resp. $\bar{A}_x$) is the space of the forms of A (resp. $\bar{A}$) at x.

To establish the relation between the definitions let x^α, $1 \leqq \alpha, \beta \leqq n$, be a local coordinate system. Then a basis in the tangent space T_x is given by $\frac{\partial}{\partial x^\alpha}$ and its dual basis T_x^* consists of the differential forms dx^β. The endomorphism J_x will be defined by

$$J_x\left(\sum_\alpha \xi^\alpha \frac{\partial}{\partial x^\alpha}\right) = \sum_{\alpha,\beta} a_\beta^\alpha \xi^\beta \frac{\partial}{\partial x^\alpha} \tag{3.2}$$

The condition that $J_x^2 = -1_x$ is expressed by

$$\sum_\beta a_\beta^\alpha a_\gamma^\beta = - \delta_\gamma^\alpha , \quad 1 \leqq \alpha,\beta,\gamma \leqq n . \tag{3.3}$$

At each point $x \in M$ the discussions of §2 apply, and we see that the forms

$$\sum_\beta \left(a_\beta^\alpha + i\delta_\beta^\alpha\right) dx^\beta \tag{3.4}$$

are of type (1,0). They are n in number and exactly $m = n/2$ of them are linearly independent over the ring of complex-valued C^∞-functions (cf. (2.3)). (The situation being local, we restrict ourselves to a sufficiently small neighborhood. As all our functions are C^∞ unless otherwise specified, we will later on frequently omit the adjective "C^∞".)

(B) A complex manifold has an almost complex structure.

In fact, the complex-valued 1-forms which, in terms of the local coordinates z^k, $1 \leqq k \leqq m$, are linear combinations of dz^k, are well-defined in a complex manifold M. These we define to be the forms of type (1,0). Since

$$\left(\frac{i}{2}\right)^m \bigwedge_k dz^k \wedge d\bar{z}^k \neq 0 ,$$

we have defined an almost complex structure on M.

To describe J in terms of the local coordinates z^k let

$$z^k = x^k + iy^k . \tag{3.5}$$

Then we have, using the fact that dz^k is of type (1,0),

$$(dz^k)\left(\frac{\partial}{\partial x^j}\right) = \delta^k_j , \quad (dz^k)\left(\frac{\partial}{\partial y^j}\right) = i\delta^k_j$$

$$(dz^k)\left(J\frac{\partial}{\partial x^j}\right) = i\delta^k_j , \quad (dz^k)\left(J\frac{\partial}{\partial y^j}\right) = -\delta^k_j , \ 1 \leqq j, k \leqq m .$$

It follows that

$$J\left(\frac{\partial}{\partial x^j}\right) = \frac{\partial}{\partial y^j} , \quad J\left(\frac{\partial}{\partial y^j}\right) = -\frac{\partial}{\partial x^j} . \tag{3.6}$$

The question arises whether this is the only way to get an almost complex manifold, i.e., whether every almost complex manifold is complex. This is the case for $n = 2$, but not in general. The question is whether local coordinates x^k, y^k, $1 \leqq k \leqq m = n/2$, can be introduced such that, if z^k are defined by (3.5), the forms of type (1,0) are linear combinations of dz^k. Suppose the almost complex structure is locally defined by the forms θ^k of type (1,0) which are linearly independent (over the ring of complex-valued C^∞-functions). Their exterior derivatives can be written

$$(3.7)\quad d\theta^k = 1/2 \sum_{j,\ell} A^k_{j\ell}\, \theta^j \wedge \theta^\ell + \sum_{j,\ell} B^k_{j\ell}\, \theta^j \wedge \bar\theta^\ell + 1/2 \sum_{j,\ell} C^k_{j\ell}\, \bar\theta^j \wedge \bar\theta^\ell$$

where $A^k_{j\ell}$, $B^k_{j\ell}$, $C^k_{j\ell}$ are complex-valued functions satisfying

$$(3.8)\quad A^k_{j\ell} + A^k_{\ell j} = C^k_{j\ell} + C^k_{\ell j} = 0, \quad 1 \leqq j,k,\ell \leqq m .$$

The condition

$$(3.9)\quad d\theta^k \equiv 0, \quad \text{mod } \theta^j$$

remains invariant under a linear transformation of the θ^k. It is satisfied if $\theta^k = dz^k$. Thus it is a necessary condition for an almost complex structure to arise from a complex structure. We will call (3.9) the *integrability condition*. By (3.7) it can also be written

$$(3.9a)\quad C^k_{j\ell} = 0 .$$

Before proceeding, we will express the integrability condition in terms of the tensor field a^α_β which defines the endomorphism J_x. Suppose that our Greek indices range from 1 to n:

$$(3.10)\quad 1 \leqq \alpha,\beta,\gamma,\lambda,\mu,\rho,\sigma \leqq n .$$

Then we have:

(C) (Eckmann-Fröhlicher) Let

$$(3.11)\quad \begin{aligned} a^\alpha_{\beta\gamma} &= -a^\alpha_{\gamma\beta} = \frac{\partial a^\alpha_\beta}{\partial x^\gamma} - \frac{\partial a^\alpha_\gamma}{\partial x^\beta}, \\ t^\alpha_{\beta\gamma} &= \sum_\rho \left(a^\alpha_{\beta\rho}\, a^\rho_\gamma - a^\alpha_{\gamma\rho}\, a^\rho_\beta \right). \end{aligned}$$

The integrability condition of the almost complex structure defined by the tensor field a^α_β is $t^\alpha_{\beta\gamma} = 0$.

Since the forms of type (1,0) are linear combinations of

those in (3.4), the integrability condition can be expressed by

$$\sum_\beta da^\alpha_\beta \wedge dx^\beta \equiv 0 \ , \quad \bmod \sum_\lambda (a^\gamma_\lambda + i\delta^\gamma_\lambda)dx^\lambda \ ,$$

or

$$\sum_{\beta,\gamma} a^\alpha_{\beta\gamma}\, dx^\beta \wedge dx^\gamma \equiv 0 \ , \quad \bmod \sum_\lambda (a^\gamma_\lambda + i\delta^\gamma_\lambda)dx^\lambda \ .$$

If we equate to zero the forms in (3.4), a fundamental system of solutions of the resulting linear equations in dx^β can be selected from $a^\gamma_\lambda - i\delta^\gamma_\lambda$ (cf. (2.1)). The condition above can therefore be written

$$\sum_{\beta,\gamma} a^\alpha_{\beta\gamma}(a^\beta_\lambda - i\delta^\beta_\lambda)(a^\gamma_\mu - i\delta^\gamma_\mu) = 0 \ .$$

Equating to zero either the real or the imaginary part of this equation, we get $t^\alpha_{\beta\gamma} = 0$.

Remark. It can be verified that $t^\alpha_{\beta\gamma}$ are the components of a tensor field.

The integrability condition is identically satisfied when $n = 2$, as can be seen from (3.9a). For $n \geqq 4$ the condition is clearly non-trivial. An almost complex structure satisfying the integrability condition is called *integrable,* otherwise *non-integrable.* An almost complex manifold of dimension $\geqq 4$ always has a non-integrable almost complex structure, for even if the given one is integrable, it can be perturbed slightly to give a non-integrable one.

A significant example of an almost complex manifold is the 6-sphere S^6 . From the theory of Lie groups it is known that S^6 can be considered as a coset space $G_2/SU(3)$, where G_2 is the exceptional simple Lie group of 14 dimensions and SU(3) is the special unitary group in three variables. From the definition of G_2 and its structure equations one sees immediately that S^6 has a non-integrable almost complex structure.

Suppose that we have an integrable almost complex structure. The condition (3.9) suggests us to apply the theorem of Frobenius on completely integrable differential systems. Since the forms are complex-valued, it will be necessary to suppose that the almost complex structure is real analytic, i.e., that the functions a^{α}_{β} are real analytic. Under this hypothesis it follows from Frobenius's theorem that there exist complex local coordinates z^k such that the forms of type (1,0) are linear combinations of dz^k . In a neighborhood where two such local coordinate systems z^k and w^j are both valid dw^j are linear combinations of dz^k , which implies that w^j are holomorphic functions of z^k . Thus the manifold has a complex structure.

This theorem that a complex structure can be introduced in a manifold with an integrable almost complex structure is also true if the latter is C^{∞} or satisfies even weaker smoothness conditions. This was first proved by A. Newlander and L. Nirenberg [20]. Subsequent proofs were given by A. Nijenhuis and W. B. Woolf, J. Kohn and L. Hörmander. These proofs are rather difficult. The case $n = 2$ is a classical theorem of Korn and Lichtenstein which asserts that a two-dimensional riemannian metric of class $C^{1,\alpha}$ $(0 < \alpha < 1)$ is locally conformal to a flat metric. Even the proof of the Korn-Lichtenstein theorem is not simple [16].

Thus we see that integrable almost complex structures and complex structures are essentially identical. In some of the problems it is not necessary to make use of the local complex coordinates z^k , and the Newlander-Nirenberg theorem will not be needed. But we will not insist on this point.

The integrability condition (3.9) or $t^{\alpha}_{\beta\gamma} = 0$ (by (C)) is a criterion for deciding whether a given almost complex structure is integrable. It gives no information on the problem whether an almost complex manifold can be given a complex structure, whose underlying almost complex structure may be different from the given one. Recently van de Wen gave examples of compact four-dimensional almost complex manifolds which do not have any complex structure; his proof makes use of the Atiyah-Singer index theorem [21]. It is an outstanding problem whether S^6 can have a complex structure.

Let M be an almost complex manifold of dimension $n = 2m$. All complex-valued C^∞-forms of type (p,q) constitute a module A_{pq} over the ring of complex valued C^∞-functions. The following properties are easily verified:

(1) If $\alpha \in A_{pq}$, then $\bar{\alpha} \in A_{qp}$.

(2) If $\alpha \in A_{pq}$, $\beta \in A_{rs}$, then $\alpha \wedge \beta \in A_{p+r,q+s}$.

(3) $dA_{pq} \subset A_{p+2,q-1} + A_{p+1,q} + A_{p,q+1} + A_{p-1,q+2}$.

(4) $A_{pq} = 0$ if p or $q > m$.

From (3) we define, for $\alpha \in A_{pq}$, the operators

$$\partial\alpha = \Pi_{p+1,q}\, d\alpha, \quad \bar{\partial}\alpha = \Pi_{p,q+1}\, d\alpha. \tag{3.12}$$

If the almost complex structure is integrable, (3) becomes

(3I) $$dA_{pq} \subset A_{p+1,q} + A_{p,q+1},$$

as follows immediately from (3.7). We can then write

$$d = \partial + \bar{\partial}. \tag{3.13}$$

Since $d^2 = 0$, we get

$$\partial^2 + \partial\bar{\partial} + \bar{\partial}\partial + \bar{\partial}^2 = 0.$$

Equating to zero the terms of different types, we find

$$\partial^2 = \partial\bar{\partial} + \bar{\partial}\partial = \bar{\partial}^2 = 0. \tag{3.14}$$

The last condition gives rise to the following form of the integrability condition:

(D) An almost complex structure is integrable if and only if

$$\bar{\partial}^2 = 0 .$$

It remains to prove that the integrability condition is satisfied if $\bar{\partial}^2 = 0$. In fact, let F be a complex-valued C^∞-function. We write

$$dF = \sum_k F_k \theta^k + \sum_k G_k \bar{\theta}^k$$

Then we have

$$\partial F = \sum_k F_k \theta^k , \quad \bar{\partial} F = \sum_k G_k \bar{\theta}^k ,$$

and

$$\bar{\partial}^2 F = \Pi_{0,2} \, d\bar{\partial} F = \Pi_{0,2} \, d(\bar{\partial} - d)F = - \Pi_{0,2} \, d\partial F$$

$$= - \sum_{k,j,\ell} F_k C^k_{j\ell} \, \bar{\theta}^j \wedge \bar{\theta}^\ell .$$

Since this expression is zero for any F, we get $C^k_{j\ell} = 0$, which is the integrability condition (3.9a).

From now on suppose M is a complex manifold. A form $\alpha \in A_{pq}$ is called $\bar{\partial}$-closed if $\bar{\partial}\alpha = 0$. Let C_{pq} be the space of $\bar{\partial}$-closed forms of type (p,q). The quotient groups

(3.15) $$D_{pq}(M) = C_{pq}/\bar{\partial} A_{p,q-1}$$

are called the *Dolbeault groups* of M.

The Dolbeault groups are analogous to the de Rham groups of a real manifold, whose definitions we recall as follows: Let A_r be the space of real-valued C^∞-forms of degree r, and C_r be the subspace of the forms of A_r which are annihilated by d. Then the de Rham groups are

(3.16) $$R_r(M) = C_r/dA_{r-1} .$$

Both the de Rham groups and the Dolbeault groups are isomorphic to cohomology groups with coefficient sheaves, which will be treated in §4. Before concluding this section, we will prove an important lemma:

(E) (The Dolbeault-Grothendieck Lemma)

In the number space C_m with the coordinates z^k, $1 \leq k \leq m$, let D be the polydisc $|z^k| < r^k$, and let D' be the smaller polydisc $|z^k| < r'^k$, $r'^k < r^k$. Let α be a form of type (p,q), $q \geq 1$, in D such that $\bar{\partial}\alpha = 0$. There exists a form β of type $(p,q-1)$ in D such that $\bar{\partial}\beta = \alpha$ in D'.

We consider first a special case of this lemma, i.e., $m = 1$, $(p,q) = (0,1)$. We write z for z^1. Then

$$\alpha = f(z)d\bar{z} ,$$

where $f(z)$ is a complex-valued C^∞-function. The form β sought is a function which satisfies the partial differential equation

(3.17) $$\frac{\partial\beta}{\partial\bar{z}} = f(z) ,$$

where

(3.18) $$\frac{\partial}{\partial\bar{z}} = 1/2 \left(\frac{\partial}{\partial x} + i \frac{\partial}{\partial y}\right) , \quad z = x + iy .$$

We note that if the equation (3.17) is split into its real and imaginary parts we get an elliptic system of two equations of the first order in two independent and two dependent variables.

Let $z, \zeta \in D$ and regard z to be fixed. We have the relation

$$d\left(\frac{\beta d\zeta}{\zeta - z}\right) = \beta_{\bar{\zeta}} \frac{d\bar{\zeta} \wedge d\zeta}{\zeta - z} .$$

Suppose $z \in D'$ and let Δ_ε be a disc of radius ε about z, ε being sufficiently small. Applying Stokes' theorem to the domain

$D' - \Delta_\varepsilon$, we get

$$\int_{\partial D'} \frac{\beta(\zeta)d\zeta}{\zeta - z} - \int_{\partial\Delta_\varepsilon} \frac{\beta(\zeta)d\zeta}{\zeta - z} = \int_{D'-\Delta_\varepsilon} \frac{\beta_{\bar\zeta}d\bar\zeta \wedge d\zeta}{\zeta - z}$$

The second integral at the left-hand side tends to $2\pi i\beta(z)$ as $\varepsilon \to 0$. We have therefore the generalized Cauchy integral formula

$$(3.19) \qquad 2\pi i\beta(z) = \int_{\partial D'} \frac{\beta d\zeta}{\zeta - z} + \int_{D'} \frac{\beta_{\bar\zeta}d\zeta \wedge d\bar\zeta}{\zeta - z}$$

Taking the conjugate complex of this equation and replacing $\bar\beta$ by β , we have also

$$(3.19a) \qquad -2\pi i\beta(z) = \int_{\partial D'} \frac{\beta d\bar\zeta}{\bar\zeta - \bar z} - \int_{D'} \frac{\beta_\zeta d\zeta \wedge d\bar\zeta}{\bar\zeta - \bar z} .$$

Equation (3.19) shows that if (3.17) has a solution $\beta(z)$, it is given by

$$(3.20) \qquad 2\pi i\beta(z) = \int_{D'} \frac{f(\zeta)d\zeta \wedge d\bar\zeta}{\zeta - z} + g(z)$$

where $g(z)$ is a holomorphic function. It remains to verify that the function in (3.20) satisfies the equation (3.17).

For this purpose we consider the relation

$$d\{f(\zeta)\log|\zeta-z|^2 d\bar\zeta\} = f_\zeta \log|\zeta-z|^2 d\zeta \wedge d\bar\zeta + \frac{f}{\zeta-z} d\zeta \wedge d\bar\zeta ,$$

and apply Stokes' theorem to the domain $D' - \Delta_\varepsilon$. As $\varepsilon \to 0$, the integral

$$\int_{\partial\Delta_\varepsilon} f(\zeta)\log|\zeta-z|^2 d\bar\zeta$$

tends to zero, because, if $|f(\zeta)| \leqq B$, we have

$$\left| \int_{\partial\Delta_\varepsilon} f(\zeta)\log|\zeta-z|^2 d\bar{\zeta} \right| \leq 4\pi B\, \varepsilon \log \varepsilon .$$

We have therefore

$$\int_{\partial D'} f(\zeta)\log|\zeta-z|^2 d\bar{\zeta} \;-\; \int_{D'} f_\zeta \log|\zeta-z|^2 d\zeta \wedge d\bar{\zeta}$$

$$= \int_{D'} \frac{f(\zeta)}{\zeta-z}\, d\zeta \wedge d\bar{\zeta} \;=\; 2\pi i\beta(z) - g(z) ,$$

by (3.20). Differentiating under the integral sign with respect to $\bar{z}$, we get

$$- \int_{\partial D'} \frac{f(\zeta)}{\bar{\zeta}-\bar{z}}\, d\bar{\zeta} \;+\; \int_{D'} f_\zeta \frac{d\zeta \wedge d\bar{\zeta}}{\bar{\zeta} - \bar{z}} \;=\; 2\pi i \,\frac{\partial\beta}{\partial\bar{z}} .$$

This differentiation can be justified, essentially because the resulting integrals exist. By (3.19a) (with β replaced by f) we see that the function $\beta(z)$ in (3.20) satisfies the equation (3.17).

It is important to remark that the proof shows that if the function $f(z)$ is holomorphic in some complex parameters, the same is true for the solution β.

To prove the general case we introduce the <u>hypothesis</u> (H_j): α does not contain $d\bar{z}^{j+1},\ldots,d\bar{z}^m$. We shall prove that if the lemma is true with the additional hypothesis (H_{j-1}), it is true with the additional hypothesis (H_j). Under the hypothesis (H_0), we have $\alpha = 0$, and the lemma is true. On the other hand, the hypothesis (H_m) is empty. Thus the above induction statement will imply the lemma.

Suppose therefore that the lemma is true with the additional hypothesis (H_{j-1}). If α does not involve $d\bar{z}^{j+1},\ldots,d\bar{z}^m$, we write

$$\alpha = \left(d\bar{z}^j \wedge \lambda\right) + \mu ,$$

where λ and μ are forms of types $(p,q-1)$ and (p,q) respectively and do not contain $d\bar{z}^j,\ldots,d\bar{z}^m$. Since $\bar{\partial}\alpha = 0$, their coefficients are holomorphic in $z^{j+1},\ldots,z^m$. By the special case proved above, we can find a form λ' of type $(p,q-1)$ which satisfies the equation

$$\frac{\partial}{\partial \bar{z}^j}\lambda' = \lambda$$

in D' and whose coefficients are holomorphic in $z^{j+1},\ldots,z^m$; here the operator $\partial/\partial\bar{z}^j$ means the operator applied to each of the coefficients. Then $\bar{\partial}\lambda' - d\bar{z}^j \wedge \lambda = \nu$ (say) does not contain $d\bar{z}^j,\ldots,d\bar{z}^m$, and

$$\alpha = \bar{\partial}\lambda' + \mu - \nu .$$

Since $\bar{\partial}\alpha = 0$, we have $\bar{\partial}(\mu-\nu) = 0$. But $\mu - \nu$ does not contain $d\bar{z}^j,\ldots,d\bar{z}^m$, so that, by our induction hypothesis we can find a form ρ of type $(p,q-1)$ in D satisfying

$$\mu - \nu = \bar{\partial}\rho \quad \text{in } D' .$$

Thus $\alpha = \bar{\partial}(\lambda' + \rho)$ and the induction is complete.

§4. Sheaves and Cohomology

Sheaf theory is a basic tool in the study of complex manifolds. We will review its main ideas and the cohomology theory built on it. For details cf. [5] or [2].

Let M be a topological space. A sheaf of abelian groups is a topological space S together with mapping $\pi\colon S \to M$, such that the following conditions are satisfied:

(1) π is a local homeomorphism;

(2) for each point $x \in M$ the set $\pi^{-1}(x)$ (called the stalk over x) has the structure of an abelian group;

(3) the group operations are continuous in the topology of S.

Let U be an open set of M . A section of the sheaf S over U is a continuous mapping $f: U \to S$, such that $\pi \circ f$ = identity. The set $\Gamma(U,S)$ of all the sections over U forms an abelian group, for if $f,g \in \Gamma(U,S)$, we can define $f-g$ by $(f-g)(x) = f(x) - g(x)$, $x \in U$. The zero of the group $\Gamma(U,S)$ is the zero section which assigns the zero of the stalk $\pi^{-1}(x)$ to every $x \in U$.

If V is an open subset of U , there is a homomorphism $\rho_{VU}: \Gamma(U,S) \to \Gamma(V,S)$ defined by restriction. These conditions motivate the following definition:

A presheaf of abelian groups over M consists of:

(1) a basis for the open sets of M ;

(2) an abelian group S_U assigned to each open set U of the basis; and

(3) a homomorphisn $\rho_{VU}: S_U \to S_V$ associated to each inclusion $V \subset U$, such that $\rho_{WV}\rho_{VU} = \rho_{WU}$ whenever $W \subset V \subset U$.

From the presheaf one can construct the sheaf by a limit process.

Suppose now that M is a complex manifold. The following sheaves will play an important rôle in future discussions:

(1) the sheaf A_{pq} of germs of complex-valued C^∞ forms of type (p,q). In particular, we will write $A = A_{00}$, the sheaf of germs of complex-valued C^∞-functions.

(2) the sheaf C_{pq} of germs of complex valued C^∞ forms of type (p,q) , which are closed under $\bar{\partial}$. We write $O = C_{00}$, the sheaf of germs of holomorphic functions. For complex manifolds this is the most important sheaf.

(3) the sheaf O^* of germs of holomorphic functions which vanish nowhere. Here the group operation is the multi-

plication of germs of holomorphic functions.

A section of the sheaf $\mathcal{A}_{pq}$ is a form of type (p,q), etc. Thus, in the notation of §3,

$$A_{pq} = \Gamma(M,\mathcal{A}_{pq}), \quad C_{pq} = \Gamma(M,\mathcal{C}_{pq}), \text{ etc.} \tag{4.1}$$

Let

$$\pi: \mathcal{S} \to M, \quad \tau: \mathcal{T} \to M$$

be two sheaves of abelian groups over the same space M. A sheaf mapping $\phi: \mathcal{S} \to \mathcal{T}$ is a continuous mapping such that $\pi = \tau \circ \phi$, i.e., a mapping which preserves the stalks: $\phi(\pi^{-1}(x)) \subset \tau^{-1}(x)$. ϕ is called a *sheaf homomorphism* if its restriction to every stalk is a homomorphism of the groups.

If $\mathcal{Q} \to M$ is a third sheaf over M, the sequence of sheaves

$$0 \to \mathcal{S} \xrightarrow{i} \mathcal{T} \xrightarrow{\phi} \mathcal{Q} \to 0 \tag{4.2}$$

connected by homomorphisms is called an *exact sequence* if at each stage the kernel of one homomorphism is identical to the image of the preceding homomorphism. We describe this by saying that $\mathcal{S}$ is a *sub-sheaf* of $\mathcal{T}$ and $\mathcal{Q}$ is the *quotient sheaf* of $\mathcal{T}$ by $\mathcal{S}$.

It follows from the Dolbeault-Grothendieck lemma proved in §3 that the sequence

$$0 \to \mathcal{C}_{pq} \xrightarrow{i} \mathcal{A}_{pq} \xrightarrow{\bar\partial} \mathcal{C}_{p,q+1} \to 0 \tag{4.3}$$

is exact. Here i is the inclusion homomorphism and $\bar\partial$ is the homomorphism on sheaves induced by the $\bar\partial$-operator. The Dolbeault-Grothendieck lemma says that $\bar\partial$ is onto; the exactness of the sequence at the other stages is obvious.

To develop the cohomology theory with a coefficient sheaf we suppose that M is a paracompact Hausdorff space. Let $\mathcal{U} = \{U_i\}$ be a locally finite open covering of M. The nerve $N(\mathcal{U})$ of the covering $\mathcal{U}$ is a simplicial complex whose vertices are the members U_i of the covering such that $U_{i_0}, U_{i_1}, \ldots, U_{i_q}$ span a q-dimensional simplex if and only if the intersection $U_{i_0} \cap U_{i_1} \cap \ldots \cap U_{i_q} \neq \emptyset$. Let $\pi: S \to M$ be a sheaf of abelian groups over M. A q-cochain of $N(\mathcal{U})$ with coefficients in the sheaf S is a function f which associates to each q-simplex $\sigma = U_{i_0} U_{i_1} \ldots U_{i_q} \in N(\mathcal{U})$ a section $f(\sigma) \in \Gamma(U_{i_0} \cap \ldots \cap U_{i_q}, S)$. Since the set of sections is an abelian group, the set of all q-cochains form an abelian group $C^q(N(\mathcal{U}),S)$.

A coboundary operator

$$\delta_q: C^q(N(\mathcal{U}),S) \to C^{q+1}(N(\mathcal{U}),S)$$

is defined as follows: if $f \in C^q(N(\mathcal{U}),S)$ and $\sigma = U_0 \ldots U_{q+1}$, then $\delta_q f \in C^{q+1}(N(\mathcal{U}),S)$ has for σ the value

$$(4.4) \qquad (\delta_q f)(\sigma) = \sum_{j=0}^{q+1} (-1)^j \rho_\sigma f(U_0 \ldots U_{j-1} U_{j+1} \ldots U_{q+1}),$$

where ρ_σ denotes the restriction of the sections to the open set $U_0 \cap \ldots \cap U_{q+1}$.

It is immediately verified that

$$(4.5) \qquad \delta_{q+1}\delta_q = 0, \quad q \geq 0.$$

The kernel of δ_q is called the group of all q-cocycles and will be denoted by $Z^q(N(\mathcal{U}),S)$. The image of δ_{q-1} is called the group of all q-coboundaries and will be denoted by $B^q(N(\mathcal{U}),S)$. As a consequence of (4.5), a q-coboundary is a q-cocycle, and the quotient group

$$(4.6) \qquad H^q(N(\mathcal{U}),S) = Z^q(N(\mathcal{U}),S)/B^q(N(\mathcal{U}),S), \ B^0 = 0,$$

is called the *q-th cohomology group of the nerve* $N(\mathcal{U})$ *with the coefficient sheaf* S.

The zeroth cohomology group has the simple interpretation:

$$H^0(N(\mathcal{U}),\mathcal{S}) = \Gamma(M,\mathcal{S}) . \tag{4.7}$$

By a standard process initiated by Čech, one can pass from the cohomology groups $H^q(N(\mathcal{U}),\mathcal{S})$ relative to all the locally finite open coverings $\mathcal{U}$ of M, to the cohomology groups $H^q(M,\mathcal{S})$, $q \geqq 0$, of the space M itself.

Let $\pi: \mathcal{S} \to M$ be a sheaf of abelian groups over M and let $\mathcal{U} = \{U_i\}$ be a locally finite open covering of M. A *partition of unity* of the sheaf $\mathcal{S}$ subordinate to the covering $\mathcal{U}$ is a collection of sheaf homomorphisms $\eta_i: \mathcal{S} \to \mathcal{S}$ with the properties:

(1) η_i is the zero map in an open neighborhood of $M - U_i$;

(2) $\Sigma_i \eta_i = 1$, the latter being the identity mapping of the sheaf $\mathcal{S}$.

A sheaf $\mathcal{S}$ of abelian groups is *fine* if it admits a partition of unity subordinate to any locally finite open covering.

Examples of fine sheaves are $\mathcal{A}_{pq}$. Examples of sheaves which are generally not fine include:

(1) the constant sheaf;

(2) the sheaf $\mathcal{C}_{pq}$.

Fine sheaves play a catalytic rôle in the cohomology theory of sheaves, because of the theorem:

If $\mathcal{S}$ is fine, then $H^q(M,\mathcal{S}) = 0$, $q \geqq 1$.

A sheaf homomorphism $i: \mathcal{S} \to \mathcal{T}$ induces a homomorphism $\Gamma(U,\mathcal{S}) \to \Gamma(U,\mathcal{T})$ for every open set U of M, and hence a homomorphism

$$i^q: C^q(N(\mathcal{U}),\mathcal{S}) \to C^q(N(\mathcal{U}),\mathcal{T}) .$$

This leads to an induced homomorphism

$$i^q: H^q(M,\mathcal{S}) \to H^q(M,\mathcal{T}) , \quad q \geqq 0 .$$

As a result of the exact sequence (4.2) we wish to describe a homomorphism

$$\delta^q\colon H^q(M,\mathcal{Q}) \to H^{q+1}(M,\mathcal{S})$$

and to connect the homomorphisms into a long exact sequence. The exact sequence (4.2) induces the exact sequence

$$0 \to C^q(N(\mathcal{U}),\mathcal{S}) \xrightarrow{i^q} C^q(N(\mathcal{N}),\mathcal{T}) \xrightarrow{\phi^q} C^q(N(\mathcal{U}),\mathcal{Q}) .$$

We put

$$\bar{C}^q(N(\mathcal{U}),\mathcal{Q}) = \phi^q\, C^q(N(\mathcal{U}),\mathcal{T}) \subset C^q(N(\mathcal{U}),\mathcal{Q}) , \tag{4.8a}$$

so that the sequence

$$0 \to C^q(N(\mathcal{U}),\mathcal{S}) \xrightarrow{i^q} C^q(N(\mathcal{U}),\mathcal{T}) \xrightarrow{\phi^q} \bar{C}^q(N(\mathcal{U}),\mathcal{Q}) \to 0$$

is exact. Let

$$\begin{aligned} \bar{Z}^q(N(\mathcal{U}),\mathcal{Q}) &= \{f \in \bar{C}^q(N(\mathcal{U}),\mathcal{Q}) \mid \delta_q f = 0\} \\ \bar{H}^q(N(\mathcal{U}),\mathcal{Q}) &= \bar{Z}^q(N(\mathcal{U}),\mathcal{Q})/\delta_{q-1}\, \bar{C}^{q-1}(N(\mathcal{U}),\mathcal{Q}) . \end{aligned} \tag{4.8b}$$

Consider the diagram

$$\begin{array}{ccccccccc}
 & & \downarrow & & \downarrow & & \downarrow & & \\
0 & \to & C^q(N(\mathcal{U}),\mathcal{S}) & \xrightarrow{i^q} & C^q(N(\mathcal{U}),\mathcal{T}) & \xrightarrow{\phi^q} & \bar{C}^q(N(\mathcal{U}),\mathcal{Q}) & \to & 0 \\
 & & \delta^q\downarrow & & \delta^q\downarrow & & \delta^q\downarrow & & \\
0 & \to & C^{q+1}(N(\mathcal{U}),\mathcal{S}) & \xrightarrow{i^{q+1}} & C^{q+1}(N(\mathcal{U}),\mathcal{T}) & \xrightarrow{\phi^{q+1}} & \bar{C}^{q+1}(N(\mathcal{U}),\mathcal{Q}) & \to & 0 \\
 & & \delta^{q+1}\downarrow & & \delta^{q+1}\downarrow & & \delta^{q+1}\downarrow & & \\
0 & \to & C^{q+2}(N(\mathcal{U}),\mathcal{S}) & \xrightarrow{i^{q+2}} & C^{q+2}(N(\mathcal{U}),\mathcal{T}) & \xrightarrow{\phi^{q+2}} & \bar{C}^{q+2}(N(\mathcal{U}),\mathcal{Q}) & \to & 0 \\
 & & \downarrow & & \downarrow & & \downarrow & &
\end{array}$$

This diagram is commutative, in the sense that the image of a cochain depends only on its final position and is independent of the paths taken. Moreover, the horizontal sequences are exact. To an element of $\bar{H}^q(M,\mathcal{Q})$ we take a representative q-cocycle, i.e., an element $u \in \bar{C}^q(N(\mathcal{U}),\mathcal{Q})$, such that $\delta^q u = 0$. There exists $v \in C^q(N(\mathcal{U}),\mathcal{T})$, such that $\phi^q v = u$. Then $\phi^{q+1} \circ \delta^q v = \delta^q \circ \phi^q v = \delta^q u = 0$, and there exists $w \in C^{q+1}(N(\mathcal{N}),\mathcal{S})$, satisfying $i^{q+1} w = \delta^q v$. w is a cocycle, for

$$i^{q+2} \circ \delta^{q+1} w = \delta^{q+1} \circ i^{q+1} w = \delta^{q+1} \circ \delta^q v = 0 ,$$

so that $\delta^{q+1} w = 0$. By further "chasing" of the diagram, it can be shown that the element of $H^{q+1}(N(\mathcal{U}),\mathcal{S})$ defined by w is independent of the various choices made. This defines a homomorphism

$$\delta^q\colon \bar{H}^q(N(\mathcal{U}),\mathcal{Q}) \to H^{q+1}(N(\mathcal{U}),\mathcal{S}) .$$

This definition is valid for a general topological space M. It can be proved that if M is Hausdorff and paracompact, then

$$\bar{H}^q(M,\mathcal{Q}) \cong H^q(M,\mathcal{Q}) .$$

A *fundamental fact* in cohomology theory is the result: If the sequence of sheaves (4.2) is exact, the sequence of cohomology groups

$$0 \longrightarrow H^0(M,\mathcal{S}) \xrightarrow{i^0} H^0(M,\mathcal{T}) \xrightarrow{\phi^0} H^0(M,\mathcal{Q}) \xrightarrow{\delta^0} H^1(M,\mathcal{S})$$

(4.9)

$$\xrightarrow{i^1} H^1(M,\mathcal{T}) \xrightarrow{\phi^1} H^1(M,\mathcal{Q}) \xrightarrow{\delta^1} H^2(M,\mathcal{S}) \longrightarrow \cdots$$

is exact.

We apply this result to the exact sequence (4.3). A section of the induced sequence of cohomology groups will be

(4.10) $$\cdots \to H^{r-1}(M,A_{pq}) \to H^{r-1}(M,C_{p,q+1}) \to H^r(M,C_{pq}) \to H^r(M,A_{pq}) \to \cdots .$$

Since the sheaf A_{pq} is fine, we have

$$H^r(M,A_{pq}) = 0 , \qquad r \geqq 1$$

and it follows from the exactness of (4.10) that we have the isomorphisms

$$(4.11)\qquad \begin{aligned} H^r(M,C_{pq}) &\cong H^{r-1}(M,C_{p,q+1}) \cong \dots \cong H^1(M,C_{p,q+r-1}) \\ &\cong H^0(M,C_{p,q+r})/\bar\partial H^0(M,A_{p,q+r-1}) . \end{aligned}$$

Comparing with (4.1), we see that the latter is the Dolbeault group $D_{p,q+r}(M)$. By changing notation, we get

$$(4.12)\qquad D_{pq}(M) \cong H^q(M,C_{p0}) .$$

This gives a sheaf-theoretic interpretation of the Dolbeault groups. Notice that C_{p0} is the sheaf of germs of forms of type (p,0) with holomorphic coefficients, and, in particular, $C_{00} = \mathcal{O}$.

The sequences (4.3) can be combined into one sequence

$$(4.13)\qquad 0 \to C_{p0} \xrightarrow{i} A_{p0} \xrightarrow{\bar\partial} A_{p1} \xrightarrow{\bar\partial} \dots \xrightarrow{\bar\partial} A_{pq} \to \dots ,$$

where i is inclusion and $\bar\partial$ is defined by the $\bar\partial$-operator. The Dolbeault-Grothendieck lemma says that the sequence (4.13) is exact; the subsheaf of A_{pq} which is the image of the preceding homomorphism and the kernel of the next one is precisely C_{pq} . Since A_{pq} is fine, (4.13) is called a *fine resolution of the sheaf* C_{p0} .

A similar, but simpler, situation prevails in the case of a real differentiable manifold M . Let A^r be the sheaf of germs of C^∞ real-valued differential forms of degree r , and let C^r be the subsheaf of A^r consisting of germs of closed r-forms. Then the sequence

$$(4.14)\qquad 0 \to R \xrightarrow{i} A^0 \xrightarrow{d} A^1 \to \dots \xrightarrow{d} A^r \to \dots ,$$

where R is the constant sheaf of real numbers and i is inclusion, is exact. (4.14) is a fine resolution of the sheaf R . From the exactness of (4.14) follows the de Rham isomorphism

$$(4.15) \qquad R_r(M) \cong H^r(M; R) ,$$

where the left-hand side is the r-dimensional de Rham group of M (cf. (3.16)).

The sheaf theory discussed above can be extended to other algebraic structures, such as sheaf of rings, sheaf of modules, etc. Moreover, the group operation on a stalk may not be abelian, in which case, however, there will not be a cohomology theory.

§5. Complex Vector Bundles; Connections

Throughout this section we will denote by M a C^∞ differentiable manifold, and we will develop the properties of complex vector bundles over M. For economy the adjective "complex" is sometimes omitted.

Let

$$F = C_q = \underbrace{C \times \ldots \times C}_{q}$$

be the complex vector space of complex dimension q. Suppose F is acted on to the right by $GL(q;C)$, the general linear group in q complex variables, so that $\xi \cdot g \in F$ and

$$(5.1) \qquad (\xi g)h = \xi(gh) , \quad \xi \in F, \quad g,h \in GL(g;C) .$$

A complex vector bundle E over M consists of a space E and a projection

$$(5.2) \qquad \psi: E \to M ,$$

such that the following conditions are fulfilled:

(1) Every point $x \in M$ has a neighborhood U for which there exists a homeomorphism (a "chart")

(5.3) $$\phi_U \colon U \times F \to \psi^{-1}(U) ,$$

with

(5.4) $$\psi \circ \phi_U(y,\xi) = y , \quad y \in U , \quad \xi \in F .$$

(2) In the intersection $U \cap V$ of two such neighborhoods U,V there exists a C^∞ map $g_{UV} \colon U \cap V \to GL(q;C)$, such that

(5.5) $$\phi_U(x,\xi) = \phi_V(x,\xi') , \quad x \in U \cap V ; \quad \xi,\xi' \in F ,$$

if and only if

(5.6) $$\xi g_{UV}(x) = \xi'$$

These functions g_{UV} , the so-called *transition functions,* satisfy the compatibility relations

(5.7) $$\begin{cases} g_{UV}^{-1} = g_{VU} , \\ g_{UV}g_{VW}g_{WU} = 1 \quad \text{in} \quad U \cap V \cap W . \end{cases}$$

If $q = 1$, the vector bundle is called a *line bundle.* The set $\psi^{-1}(x)$, $x \in M$, is a complex vector space of dimension q , and is called the *fiber* at x . Our assumptions are such that the complex linear structures on the fibers have a meaning.

As a consequence of this remark, operations on complex vector spaces which commute with the actions of the general linear groups can be extended to operations on bundles. Among the most important operations are:

(1) The dual bundle E^* of E . Its transition functions are ${}^t g_{UV}^{-1}$ (i.e., the transpose inverse of g_{UV} , when the latter is interpreted as a non-singular $(q \times q)$-matrix).

(2) If E' and E'' are two complex vector bundles over M with the transition functions g'_{UV} , g''_{UV} respectively,

their *direct sum* or *Whitney sum* $E' \oplus E''$ is defined by the transition functions

$$\begin{pmatrix} g'_{UV} & 0 \\ 0 & g''_{UV} \end{pmatrix}$$

Similarly, their *tensor product* $E' \otimes E''$ is defined by the transition functions $g'_{UV} \otimes g''_{UV}$. If the dimensions of the fibers of E', E'' are q', q'' respectively, the fiber dimension of $E' \oplus E''$ is $q' + q''$ and that of $E' \otimes E''$ is $q' q''$.

(3) The bundle $\mathrm{Hom}(E',E'') \cong E'^{*} \otimes E''$.

In order that the notion of a vector bundle be meaningful, it is desirable to introduce an equivalence relation which amounts to a change of the charts. Let E and E' be two vector bundles over M with the same fiber dimension q which, relative to an open covering $\{U,V,\ldots\}$ of M, are given by the charts ϕ_U, ϕ'_U and the transition functions g_{UV}, g'_{UV} respectively. They are called *equivalent* if to each U there is a C^∞-map $g_U\colon U \to GL(q;C)$, such that

$$\phi_U(x,\xi g_U) = \phi'_U(x,\xi), \quad x \in U, \quad \xi \in F. \tag{5.8}$$

In terms of the transition functions condition (5.8) implies:

$$g'_{UV} = g_U g_{UV} g_V^{-1}. \tag{5.9}$$

An immediate question is the scope of the equivalence classes of complex vector bundles over M, or, more specifically, whether there exist bundles which are (globally) not products of M with F. For $q = 1$ the answer is given by the theorem:

(A) All the C^∞ complex line bundles over a differentiable manifold M form a group which is isomorphic to $H^2(M,Z)$, the second cohomology group of M with integer coefficients.

To prove this theorem let A be the sheaf of germs of complex-valued C^∞ functions and let A^* be the sheaf of germs of nowhere zero complex-valued C^∞ functions, the latter with multiplication as the group operation. By the compatibility relations (5.7) and by (5.9) it follows that the equivalence classes of C^∞ complex line bundles are in one-one correspondence with the elements of the cohomology group $H^1(M,A^*)$. Thus all the line bundles of M form a group, and the multiplication of two line bundles is given by the tensor product. From now on we will not distinguish between a line bundle and an equivalence class of line bundles.

Consider the sequence of sheaves

$$(5.10) \qquad 0 \to Z \xrightarrow{i} A \xrightarrow{e} A^* \to 0 .$$

where i is inclusion and e is defined by

$$(5.11) \qquad e(f(x)) = \exp(2\pi i f(x)), \qquad f(x) \in A .$$

The sequence (5.10) is obviously an exact sequence. From its exactness follows the exactness of the following sequence of cohomology groups:

$$H^1(M,A) \xrightarrow{e^1} H^1(M,A^*) \xrightarrow{\delta} H^2(M,Z) \xrightarrow{i^2} H^2(M,A) .$$

Since A is a fine sheaf, the groups at both ends of this sequence are zero, and we get the isomorphism stated in the theorem.

If $E \in H^1(M,A^*)$ is a complex line bundle, $\delta E \in H^2(M,Z)$ is called its *Chern class*.

The simple conclusion in (A) is possible, because the group $GL(1;C)$ is abelian. For general q there are Chern classes

$$c_i(E) \in H^{2i}(M,Z) , \qquad 1 \leqq i \leqq q ,$$

which are the simplest invariants of a complex vector bundle, but we will postpone their discussion to a later section.

Let E be a complex vector bundle over M, and let T^* be the cotangent bundle of M. Denote by $\Gamma(E)$ and $\Gamma(T^* \otimes E)$ respectively the spaces of sections of E and of the tensor product $T^* \otimes E$ (over C). A *connection* on E is an operator

$$D: \Gamma(E) \rightarrow \Gamma(T^* \otimes E), \tag{5.12}$$

which satisfies the conditions:

$$\begin{aligned} D(\gamma_1+\gamma_2) &= D\gamma_1+D\gamma_2, \quad \gamma_1,\gamma_2 \in \Gamma(E), \\ D(f\gamma) &= df\cdot\gamma + fD\gamma, \quad \gamma \in \Gamma(E), \end{aligned} \tag{5.13}$$

where $f \in A$ (= the space of complex-valued C^∞ functions over M) and $df\cdot\gamma = df \otimes \gamma$, the tensor product here being over A.

We will first study the local properties of a connection. Let U be an open set of M, and let $s_1,\ldots,s_q$ be a *frame field* over U, i.e., q sections of the bundle E over U, such that $s_1(x),\ldots,s_q(x)$, $x \in U$, are linearly independent. Then we can write

$$Ds_i = \sum_j \omega_i^j s_j, \quad 1 \leqq i,j \leqq q, \tag{5.14}$$

where ω_i^j are complex-valued 1-forms in U. For economy of writing we will express (5.14) in matrix form. In fact, let

$$^t s = (s_1,\ldots,s_q), \quad \omega = (\omega_i^j), \tag{5.15}$$

so that s itself is a one-columned matrix of q sections. Then (5.14) can be written

$$Ds = \omega s. \tag{5.16}$$

The matrix ω completely determines the connection. In fact, any section ξ of E over U can be written

$$\xi = \sum_i \xi^i s_i, \tag{5.17}$$

where ξ^i are complex-valued C^∞-functions in U. By (5.13), we have

$$D\xi = \sum_i (d\xi^i + \sum_j \xi^j \omega_j^i) s_i . \tag{5.18}$$

We call ω the *connection matrix.*

The section ξ is called *horizontal* if

$$D\xi = 0 \tag{5.19}$$

or

$$d\xi^i + \sum_j \xi^j \omega_j^i = 0 . \tag{5.19a}$$

Equations (5.19a) are a system of total differential equations and generally do not have a solution. However, when restricted to a parametrized curve C with parameter t, they become a system of ordinary differential equations, and a solution $\xi^i(t)$ is determined by its initial values $\xi^i(t_0)$ at a given point $t = t_0$. The mapping $C \to \psi^{-1}(C)$ defined by assigning to the point $t \in C$ the vector $\gamma = \Sigma_i \xi^i(t) s_i$ is called a *lifting* of the curve C to the bundle E, and it is called a *horizontal lifting* if γ satisfies (5.19) or (5.19a). In classical language a lifting is called a *vector field* along C and a horizontal lifting is called a *parallel vector field* along C.

Let

$$s' = gs \tag{5.20}$$

be a new frame field, where g is a non-singular $(q \times q)$-matrix of complex-valued C^∞-functions in U. By (5.13), we have

$$Ds' = \omega' s' , \tag{5.21}$$

where

$$\omega' g = dg + g\omega . \tag{5.22}$$

This is an important formula, giving the effect on the connection matrix under a change of the frame field.

By taking the exterior derivative of (5.22) and using (5.22), we get

(5.23) $$\Omega' g = g\Omega ,$$

where

(5.24) $$\Omega = d\omega - \omega \wedge \omega ,$$

and Ω' is defined similarly in terms of the connection matrix ω'. Ω is a $(q\times q)$-matrix of exterior 2-forms, and is called the *curvature matrix* relative to the frame field s.

The simple transformation law (5.23) implies the following: The vanishing of Ω is a condition independent of the choice of s. A connection satisfying $\Omega = 0$ is called *flat*.

Exterior differentiation of (5.24) gives

(5.25) $$d\Omega + \Omega \wedge \omega - \omega \wedge \Omega = 0 ,$$

which is called the *Bianchi Identity*.

The example of the curvature matrix motivates the definition: Suppose there is associated to every frame field s a $(q\times q)$-matrix Φ_s of forms of degree k, such that under a change of the frame field (5.20) we have

(5.26) $$\Phi_{s'} = g\Phi_s g^{-1} .$$

Such a collection of matrices $\{\Phi_s\}$ is called a *tensorial matrix of the adjoint type*. (The name arises from the adjoint representation of the group $GL(q;C)$.) By taking the exterior derivative of (5.26) and using (5.22), we get

(5.27) $$D\Phi_{s'} = gD\Phi_s g^{-1} ,$$

where

$$D\Phi_s = d\Phi_s - \omega \wedge \Phi_s + (-1)^k \Phi_s \wedge \omega \tag{5.28}$$

and $D\Phi_{s'}$ is defined similarly with the connection matrix ω' relative to the frame field s'. Thus $D\Phi_s$ is a tensorial matrix of $(k+1)$-forms of the adjoint type. It is called the *covariant differential* of Φ_s.

The covariant differential of the curvature matrix will not lead to anything significant because the Bianchi identity (5.25) can be written

$$D\Omega = 0 . \tag{5.29}$$

Here and later we will frequently omit the subscript s, when the frame field is fixed through the discussion.

By (5.28) it can be immediately verified that

$$D^2\Phi = DD\Phi = [\Phi,\Omega] , \tag{5.30}$$

where the "commutator" is defined by

$$[\Phi,\Omega] = \Phi \wedge \Omega - \Omega \wedge \Phi . \tag{5.31}$$

We now consider a complex-valued function $P(A_1,\ldots,A_r)$, whose arguments are the $(q\times q)$-matrices A_i, $1 \leqq i \leqq r$, and which is C-linear in each of the arguments. In fact, if

$$A_i = (a_{i,\alpha\beta}) , \quad 1 \leqq i \leqq r , \quad 1 \leqq \alpha,\beta \leqq q , \tag{5.32}$$

then

$$P(A_1,\ldots,A_r) = \Sigma\, \lambda_{\alpha_1\ldots\alpha_r\beta_1\ldots\beta_r}\; a_{1\alpha_1\beta_1} \cdots a_{r\alpha_r\beta_r} , \tag{5.33}$$

where the λ's are complex numbers and the summation is over the α's and the β's from 1 to q. Such a function (or polynomial) is called *invariant*, if

$$P(gA_1g^{-1},\ldots,gA_rg^{-1}) = P(A_1,\ldots,A_r) \tag{5.34}$$

for every non-singular matrix g . It will be called *symmetric*, if its value remains unchanged on a permutation of its arguments.

Examples of symmetric invariant polynomials can be obtained as follows: Let A be a $(q\times q)$-matrix, I be the $(q\times q)$-unit matrix, and let

$$\det(I + \frac{i}{2\pi} A) = \sum_{0\leq j\leq q} \binom{q}{j} P_j(A) , \tag{5.35}$$

where $P_j(A)$ is a polynomial of degree j in the elements of A . Let $P_j(A_1,\ldots,A_j)$ be the completely polarized polynomial of $P_j(A)$, so normalized that

$$P_j(A,\ldots,A) = P_j(A) . \tag{5.36}$$

From the definition (5.35), we have

$$P_j(gAg^{-1}) = P_j(A) . \tag{5.37}$$

Since $P_j(A_1,\ldots,A_j)$ can be expressed in terms of $P_j(A)$ for different arguments A , for instance,

$$P_2(A_1,A_2) = 1/2\{P_2(A_1+A_2) - P_2(A_1) - P_2(A_2)\} ,$$

it follows that $P_j(A_1,\ldots,A_j)$ are invariant.

Suppose $P(A_1,\ldots,A_r)$ is an invariant polynomial, so that the equation (5.34) is fulfilled. Let

$$g = I + g' .$$

Then

$$g^{-1} = I - g' + \ldots ,$$

where the dots involve terms containing higher powers of the elements of g' . Substituting into (5.34) and retaining only the terms which are linear in the elements of g' , we get

$$\sum_{1\leq i\leq r} P(A_1,\ldots,g'A_i-A_ig',\ldots,A_r) = 0 , \tag{5.38}$$

for any matrix g' . This identity remains true, when $A_1,\ldots,A_r$ are matrices whose elements are differential forms (in which case P is a complex-valued form).

Suppose the elements of A_i are forms of degree d_i . It follows from (5.38) that

$$\sum_{1\leq i\leq r} (-1)^{d_1+\ldots+d_{i-1}} P(A_1,\ldots,\theta\wedge A_i,\ldots,A_r)$$

(5.39)

$$+ \sum_{1\leq i\leq r} (-1)^{d_1+\ldots+d_i+1} P(A_1,\ldots,A_i\wedge\theta,\ldots,A_r) = 0 ,$$

where θ is a $(q\times q)$-matrix of 1-forms. In fact, θ is a sum of matrices of the form $g'\alpha$, where g' is a matrix of functions and α is a one-form. Since (5.39) is linear in θ , it suffices to prove it for the case $\theta = g'\alpha$. By moving α to the front of the expressions, we see that (5.39) for the case $\theta = g'\alpha$ follows immediately from (5.38).

The invariant polynomials constitute a link between the local properties of a connection and its global properties. In fact, we say that a family of matrices $\{\Phi_s\}$ is a tensorial matrix of k-forms of the adjoint type in M , if such a matrix Φ_s is associated to every local frame field s such that the relation (5.26) holds under a change of the frame field (5.20). If $P(A_1,\ldots,A_r)$ is an invariant polynomial and A_i is a tensorial matrix of the adjoint type in M , whose elements are forms of degree d_i , $1 \leq i \leq r$, then $P(A_1,\ldots,A_r)$ is a form of degree $d_1 + \ldots + d_r$ which is *globally defined* in M . Moreover, it follows from (5.28) and (5.39) that its exterior derivative is

(5.40)

$$dP(A_1,\ldots,A_r) = \sum_{1\leq i\leq r} (-1)^{d_1+\ldots+d_{i-1}} P(A_1,\ldots,DA_i,\ldots,A_r) .$$

For the polynomials $P_j(A)$ defined in (5.35) we have therefore

(5.41) $$dP_j(\Omega) = 0 ,$$

because of the Bianchi identity (5.29). Thus $P_j(\Omega)$ is a closed form of degree $2j$ in M and defines an element of the de Rham group $R_{2j} \cong H^{2j}(M, C)$ with complex coefficients.

(B) Let $\psi: E \to M$ be a complex vector bundle with fiber dimension q. Let Ω be the curvature matrix of a connection in the bundle. Then a change of the connection modifies $P_j(\Omega)$, $1 \leqq j \leqq q$, by an additive term of the form dQ, where Q is a form of degree $2j - 1$ in M.

The following proof of (B) is due to Weil. Let ω, Ω and $\tilde{\omega}, \tilde{\Omega}$ be respectively the connection and curvature matrices of two connections relative to the same frame field s. If s and s' are related by (5.20), we have (5.22) and the corresponding relation

$$\tilde{\omega}' g = dg + g\tilde{\omega} ,$$

for the second connection. Putting

(5.42) $$\eta = \tilde{\omega} - \omega , \qquad \eta' = \tilde{\omega}' - \omega' ,$$

we get

(5.43) $$\eta' g = g\eta .$$

Thus, η, the difference of two connection matrices, is a tensorial matrix of 1-forms of the adjoint type. We put

(5.44) $$\omega_t = \omega + t\eta , \qquad 0 \leqq t \leqq 1 .$$

Then ω_t is a connection matrix depending on the parameter t, which reduces to ω and $\tilde{\omega}$ for $t = 0$ and $t = 1$ respectively.

The curvature matrix of the connection ω_t is by definition

$$(5.45) \qquad \Omega_t = d\omega_t - \omega_t \wedge \omega_t = \Omega + tD\eta - t^2\eta \wedge \eta$$

where the covariant differential is taken with respect to the connection ω .

Let $P(A_1,\ldots,A_r)$ be a symmetric invariant polynomial. Let

$$(5.46) \qquad \begin{aligned} P(A) &= P(A,\ldots,A) \\ Q(B,A) &= rP(B,\underbrace{A,\ldots,A}_{r-1}) \end{aligned}$$

Then we have

$$\frac{d}{dt} P(\Omega_t) = Q(D\eta,\Omega_t) - 2tQ(\eta \wedge \eta,\Omega_t) .$$

On the other hand, we have, from (5.45) and (5.30),

$$\begin{aligned} D\Omega_t = tD^2\eta + t^2[\eta,D\eta] &= t[\eta,\Omega] + t^2[\eta,D\eta] \\ &= t[\eta,\Omega_t] , \end{aligned}$$

so that

$$\begin{aligned} dQ(\eta,\Omega_t) &= Q(D\eta,\Omega_t) - r(r-1)P(\eta,D\Omega_t,\Omega_t,\ldots,\Omega_t) \\ &= Q(D\eta,\Omega_t) - r(r-1)tP(\eta,[\eta,\Omega_t], \Omega_t,\ldots,\Omega_t) . \end{aligned}$$

Equation (5.39) gives, with $\theta = A_1 = \eta$, $A_2 = \ldots = A_r = \Omega_t$,

$$2Q(\eta \wedge \eta,\Omega_t) - r(r-1)P(\eta,[\eta,\Omega_t],\Omega_t,\ldots,\Omega_t) = 0 .$$

Combining the last two equations, we get

$$dQ(\eta,\Omega_t) = Q(D\eta,\Omega_t) - 2tQ(\eta \wedge \eta,\Omega_t) .$$

Therefore

$$(5.47) \qquad \frac{d}{dt} P(\Omega_t) = dQ(\eta,\Omega_t) .$$

Integrating with respect to t , we get

$$(5.48) \qquad P(\tilde{\Omega}) - P(\Omega) = d\int_0^1 Q(\eta,\Omega_t)dt .$$

This proves (B).

Special cases of $P_j(\Omega)$ are:

$$(5.49)\qquad \begin{aligned} P_1(\Omega) &= \frac{i}{2\pi q} \sum_j \Omega_j^j , \\ P_2(\Omega) &= \frac{-2}{(2\pi)^2 q(q-1)} \sum_{j<k} \left(\Omega_j^j \Omega_k^k - \Omega_j^k \Omega_k^j\right) , \end{aligned}$$

where Ω_j^k , $1 \leqq j , k \leqq q$, are the elements of the curvature matrix Ω .

We will now define an hermitian structure on the bundle (5.2). We recall that an hermitian structure on a complex vector space V is a complex-valued function $H(\xi,\eta)$, $\xi,\eta \in V$, such that

$$(5.50)\qquad \begin{aligned} &(1)\quad H(\lambda_1\xi_1 + \lambda_2\xi_2,\eta) = \lambda_1 H(\xi_1,\eta) + \lambda_2 H(\xi_2,\eta) , \\ &\qquad\qquad \lambda_1,\lambda_2 \in C , \quad \xi_1,\xi_2,\eta \in V , \\ &(2)\quad \overline{H(\xi,\eta)} = H(\eta,\xi) . \end{aligned}$$

It is called positive definite if

$$(5.51)\qquad H(\xi,\xi) > 0 , \qquad \xi \neq 0 .$$

An *hermitian structure* on the complex vector bundle (5.2) is a C^∞ field of positive definite hermitian structures in the fibers of E . That is, if ξ,η are two C^∞-sections of the bundle, $H(\xi,\eta)$ is a complex-valued C^∞-function having properties corresponding to (5.50). A complex vector bundle with an hermitian structure is called an *hermitian vector bundle*. By a partition of unity argument, it can be shown that every complex vector bundle can be given an hermitian structure.

To every frame field s the hermitian structure defines an hermitian matrix

$$(5.52) \qquad H_s = {}^t\bar{H}_s = (H(s_i, s_j)), \quad 1 \leqq i, j \leqq q ,$$

and is in turn completely determined by this matrix. Under a change of frame field (5.20) this matrix is transformed according to

$$(5.53) \qquad H_{s'} = gH_s{}^t\bar{g} ,$$

where

$$H_{s'} = (H(s_i', s_j')) , \quad 1 \leqq i,j \leqq q .$$

A connection in an hermitian vector bundle is called *admissible*, if $H(\xi,\eta)$ remains constant when ξ,η are horizontal sections along arbitrary curves. Let

$$(5.54) \qquad h_{ik} = H(s_i, s_k) , \quad 1 \leqq i, j, k \leqq q$$

and let

$$\xi = \sum_i \xi^i s_i , \quad \eta = \sum_j \eta^j s_j .$$

Then

$$H(\xi,\eta) = \sum_{i,k} h_{ik}\xi^i\bar{\eta}^k .$$

The sections ξ,η being horizontal, we have (5.19a) and a similar equation for η^k. It follows that

$$dH(\xi,\eta) = \sum_{i,k} (dh_{ik} - \Sigma h_{jk}\omega_i^j - \Sigma h_{ij}\bar{\omega}_k^j)\xi^i\bar{\eta}^k .$$

Since horizontal sections along curves exist with arbitrary initial values of ξ^i, η^k, the condition for an admissible connection becomes

$$(5.55) \qquad dh_{ik} - \sum_j h_{jk}\omega_i^j - \sum_j h_{ij}\bar{\omega}_k^j = 0 ,$$

or, in matrix notation

$$(5.55a) \qquad dH = \omega H + H{}^t\bar{\omega} ,$$

where the subscript s is dropped. By an elementary extension argument, it follows from (5.55a) that an admissible connection always exists in an hermitian vector bundle. By taking the exterior derivative of (5.55a), we get

$$\Omega H + H^t\bar{\Omega} = 0 , \tag{5.56}$$

i.e., ΩH is skew-hermitian.

A frame field s of an hermitian vector bundle is called *unitary*, if $H_s = I$ (= the unit matrix). Relative to a unitary frame field, the equations (5.55a) and (5.56) become respectively

$$\omega + {}^t\bar{\omega} = 0 , \tag{5.57}$$

$$\Omega + {}^t\bar{\Omega} = 0 , \tag{5.58}$$

i.e., the connection and curvature matrices ω and Ω are both skew-hermitian.

It follows from (5.56) that for an hermitian vector bundle with an admissible connection, we have

$$\det(I + \frac{i}{2\pi}\Omega) = \det(I - \frac{i}{2\pi}\bar{\Omega}) = \overline{\det(I + \frac{i}{2\pi}\Omega)} . \tag{5.59}$$

For the coefficients $P_j(A)$ defined in (5.35) and their polarized polynomials $P_j(A_1,\ldots,A_j)$ we write

$$P_j(A_1,\ldots,A_j) = (\mathrm{Re}P_j)(A_1,\ldots,A_j) + i(\mathrm{Im}P_j)(A_1,\ldots,A_j) , \tag{5.60}$$

so that $\mathrm{Re}P_j$ and $\mathrm{Im}P_j$ are real-valued and R-linear in each of their arguments. Let

$$(\mathrm{Re}P_j)(\Omega) = (\mathrm{Re}P_j)(\Omega,\ldots,\Omega) ,$$

$$(\mathrm{Im}P_j)(\Omega) = (\mathrm{Im}P_j)(\Omega,\ldots,\Omega) .$$

Then it follows from (5.59) that

$$(\mathrm{Im}P_j)(\Omega) = 0$$

for the curvature matrix Ω of an admissible connection of an hermitian structure. It follows from Theorem (B) that for any connection the element of the de Rham group R_{2j} determined by $(\text{Im}P_j)(\Omega)$ is zero. On the other hand, we will show later that the element determined by $(\text{Re}P_j)(\Omega)$ is what is called the jth Chern class of the bundle E with real or integer coefficients.

§6. Holomorphic Vector Bundles and Line Bundles

Let M be a complex manifold of dimension m and let $\psi: E \to M$ be a complex vector bundle over M with fiber dimension q. Relative to a covering $\{U,V,\ldots\}$ of M let g_{UV} be the transition functions of E. The bundle is called *holomorphic* if all these functions g_{UV} are *holomorphic* (i.e., g_{UV}, considered as a non-singular $(q\times q)$-matrix, is a matrix of holomorphic functions in $U \cap V$). If $q = 1$, E is called a *holomorphic line bundle.*

An example of a holomorphic vector bundle over M is the tangent bundle of M. Let $z_U^1,\ldots,z_U^m$ (respectively $z_V^1,\ldots,z_V^m$) be the local coordinates in U (resp. in V). Then the tangent bundle has as transition functions the Jacobian matrices

$$j_{UV} = \frac{\partial(z_U^1,\ldots,z_U^m)}{\partial(z_V^1,\ldots,z_V^m)} . \tag{6.1}$$

Let E be a holomorphic bundle. A section γ of E over a neighborhood $U \subset M$ is *holomorphic* if the components of γ relative to a chart are holomorphic functions. A frame field $s = {}^t(s_1,\ldots,s_q)$ is holomorphic if each s_i is a holomorphic section. When s and s' are holomorphic frame fields, the matrix g in the equation (5.20) is a matrix of holomorphic functions. A connection such that the connection matrix is a matrix of 1-forms of type (1,0) relative to a holomorphic frame field will be called a *connection of type* (1,0).

Suppose an hermitian structure is defined in E . From (5.55a) it follows that it has a uniquely defined admissible connection of type (1,0). In fact, its connection matrix is

(6.2) $$\omega = \partial H \cdot H^{-1} .$$

From (6.2) we find that its curvature matrix is

(6.3) $$\Omega = -\partial\bar{\partial}H \cdot H^{-1} + \partial H \cdot H^{-1} \wedge \bar{\partial}H \cdot H^{-1} ,$$

so that Ω is of type (1,1). (A matrix of differential forms is said to be of type (p,q) if each element is a form of type (p,q).)

In case $q = 1$, the matrices in question are (1×1)-matrices:

(6.4) $$H = (h) , \quad \Omega = (\Omega) , \quad h > 0 ,$$

and (6.3) can be written

(6.5) $$\Omega = -\partial\bar{\partial} \log h .$$

Notice that for $q = 1$ a tensorial matrix of the adjoint type is a form in M , so that Ω is globally defined in M. We call $\frac{i}{2\pi}\Omega$ the *curvature form* of the connection.

On M let $\mathcal{O}$ be the sheaf of germs of holomorphic functions and $\mathcal{O}^*$ be the sheaf of germs of nowhere zero holomorphic functions, the latter with multiplication as the group operation. A meromorphic function is locally the ratio of two holomorphic functions. Let μ be the sheaf of germs of meromorphic functions. Then $\mathcal{O}^*$ is a subsheaf of μ and the quotient sheaf $\mathcal{D}$ is by definition a sheaf of germs of divisors. The latter is locally represented by a meromorphic function defined up to the multiplication by a nowhere zero holomorphic function. We have the exact sequence

(6.6) $$0 \to \mathcal{O}^* \xrightarrow{i} \mu \xrightarrow{k} \mathcal{D} \to 0 .$$

Its induced exact cohomology sequence has the part

(6.7)

$$0 \to H^0(M,\mathcal{O}^*) \xrightarrow{i^0} H^0(M,\mu) \xrightarrow{k^0} H^0(M,\mathcal{D}) \xrightarrow{\delta^0} H^1(M,\mathcal{O}^*) \to \dots .$$

From the exactness it follows that the quotient group

(6.8)
$$H^0(M,\mathcal{D})/k^0H^0(M,\mu)$$

is isomorphic to a subgroup of $H^1(M,\mathcal{O}^*)$. An element of $H^0(M,\mathcal{D})$ is called a *divisor*. Two divisors are called *linearly equivalent* if they differ from each other (multiplicatively) by a meromorphic function in M . Thus the group (6.8) is the group of divisor-classes with respect to linear equivalence. On the other hand, $H^1(M,\mathcal{O}^*)$ is the group of all holomorphic line bundles over M , the group operation being defined by tensor product. Kodaira and Spencer proved that if M is a non-singular projective variety the group (6.8) is isomorphic to $H^1(M,\mathcal{O}^*)$, [19].

We wish to study the group $H^1(M,\mathcal{O}^*)$. For this purpose we consider the exact sequence of sheaves:

(6.9)
$$0 \to Z \xrightarrow{i} \mathcal{O} \xrightarrow{e} \mathcal{O}^* \to 0 ,$$

where e is defined by

(6.10)
$$e(f(x)) = 2\pi i \exp(f(x)) , \quad f(x) \in \mathcal{O} .$$

The sequence (6.9) leads to the homomorphism

(6.11)
$$\delta^1 \colon H^1(M,\mathcal{O}^*) \to H^2(M,Z) ,$$

and we wish to describe the image of δ^1 .

Let A_R^r be the sheaf of germs of real-valued C^∞ r-forms in M and C_R^r be the subsheaf of A_R^r consisting of those germs which are closed under d , so that C_R^0 is the constant sheaf R . Then we have the exact sequence

(6.12)
$$0 \to C_R^r \xrightarrow{\ell^r} C_R^r \xrightarrow{d} C_R^{r+1} \to 0 ,$$

whose induced exact cohomology sequence is

$$(6.13) \quad \cdots \quad H^p(M,C^r_R) \xrightarrow{\ell^{p,r}} H^p(M,A^r_R) \xrightarrow{d} H^p(M,C^{r+1}_R)$$
$$\xrightarrow{\partial^{p,r}} H^{p+1}_R(M,C^r_R) \quad \cdots \ .$$

We note that A^r_R is a fine sheaf.

We now write the following diagram of cohomology groups of M connected by homomorphisms:

$$(6.14) \qquad \begin{array}{ccccccc} & & & & H^1(A^0_R) = 0 & & \\ & & & & \downarrow d & & \\ H^0(A^1_R) & \xrightarrow{d} & H^0(C^2_R) & \xrightarrow{\partial^{0,1}} & H^1(C^1_R) & \xrightarrow{\ell^{1,1}} & H^1(A^1_R) = 0 \\ & & & & \downarrow \partial^{1,0} & & \\ H^1(\mathcal{O}^*) & \xrightarrow{\delta^1} & H^2(Z) & \xrightarrow{j} & H^2(R) & & \\ & & & & \downarrow \ell^{2,0} & & \\ & & & & H^2(A^0_R) = 0 & & \end{array}$$

In this diagram the manifold M is omitted in the notation of the cohomology groups for the sake of simplicity. The vertical sequence and the top horizontal sequence are exact, being parts of the sequence (6.13). The homomorphism j in the second horizontal sequence is induced by the coefficient homomorphism $j: Z \to R$. For an hermitian line bundle $E \in H^1(M,\mathcal{O}^*)$ we wish to determine

$$(6.15) \qquad (\partial^{0,1})^{-1} \circ (\partial^{1,0})^{-1} \circ j \circ \delta^1 E \ ,$$

which is a real-valued closed 2-form in M. In fact, we wish to show that the negative of the form (6.15) is the curvature form $\frac{i}{2\pi}\Omega$, up to an additive term $d\alpha$, where α is a real-valued 1-form in M.

This "diagram chasing" is not difficult, the main point being to remember the definitions of the homomorphisms in question. Before proceeding we will give some more discussion of the hermitian structure of a holomorphic line bundle E and its curvature form. In fact, let $\mathcal{U} = \{U, V, W, \ldots\}$ be an open covering of M which is sufficiently fine. Let s_U be a holomorphic frame field over U. The fiber dimension being one, s_U is given by a nowhere zero holomorphic function in U. Let

$$h_U = H(s_U, s_U) > 0 . \tag{6.16}$$

The change of frame field in $U \cap V$ is given by

$$s_U g_{UV} = s_V , \tag{6.17}$$

where g_{UV} is a nowhere zero holomorphic function in $U \cap V$. From (6.16) and (6.17) we derive

$$h_U |g_{UV}|^2 = h_V \quad \text{in } U \cap V . \tag{6.18}$$

It follows that

$$\partial\bar{\partial} \log h_U = \partial\bar{\partial} \log h_V \quad \text{in } U \cap V , \tag{6.19}$$

which gives a verification of the remark following formula (6.5).

Suppose M is equipped with a riemannian metric and that the members of the covering $\mathcal{U}$ are convex. Then the intersection of any number of the members of the covering, if non-empty, is convex. In $U \cap V$ $(\neq \emptyset)$ we construct the holomorphic function f_{UV} satisfying

$$g_{UV} = \exp(2\pi i\, f_{UV}) . \tag{6.20}$$

In $U \cap V \cap W \neq \emptyset$ let

$$c_{UVW} = f_{UV} + f_{VW} + f_{WU} . \tag{6.21}$$

Then

$$\exp(2\pi i c_{UVW}) = g_{UV} g_{VW} g_{WU} = 1 ,$$

so that c_{UVW} is an integer. The two-cochain of the nerve $N(\mathcal{U})$ of the covering $\mathcal{U}$ defined by assigning to the simplex UVW the integer c_{UVW} is a two-cocycle and defines a representative of $\delta^1 E$ and hence of $j \circ \delta^1 E$.

Next we wish to find a representative of the element of $H^1(C^1_R)$ which is mapped by $\partial^{1,0}$ to $j \circ \delta^1 E$. This will be given by a real-valued closed 1-form in every $U \cap V \neq \emptyset$, and we see that the form $1/2\, d(f_{UV} + \bar{f}_{UV})$ has the desired property. In fact, by (6.20) and (6.18) we have

$$\begin{aligned} 1/2\, d(f_{UV} + \bar{f}_{UV}) &= \frac{1}{4\pi i} \{\partial \log g_{UV} - \bar{\partial} \log \bar{g}_{UV}\} \\ &= \frac{1}{4\pi i} (\partial - \bar{\partial}) \log |g_{UV}|^2 \\ &= \frac{1}{4\pi i} (\partial - \bar{\partial})(-\log h_U + \log h_V) \,. \end{aligned}$$

By the definition of $\partial^{0,1}$ we get a representative of (6.15) as

$$\frac{1}{4\pi i} d(\partial - \bar{\partial}) \log h_U = \frac{i}{2\pi} \partial\bar{\partial} \log h_U = -\frac{i}{2\pi} \Omega \,.$$

Thus we have the theorem:

(A) Let $E \in H^1(M, \mathcal{O}^*)$ be a holomorphic line bundle over a complex manifold M . Let $c(E) = -\delta^1 E \in H^2(M,Z)$ be its Chern class. Suppose that E has an hermitian structure with the curvature form $\frac{i}{2\pi} \Omega$. Then the element of the de Rham group $R_2(M)$ defined by $\frac{i}{2\pi} \Omega$ corresponds to the element $jc(E) \in H^2(M,R)$ via the de Rham isomorphism (4.15), j being induced by the coefficient homomorphism $j\colon Z \to R$.

Consider the de Rham isomorphism

$$R_2(M) \xrightarrow{\ \rho\ } H^2(M,R) \,. \tag{4.15}$$

There is a subgroup $R_{11}(M)$ of $R_2(M)$, whose elements have as representatives forms of type (1,1). (Recall that an element γ of $R_2(M)$ is a class of forms $\alpha + d\beta$, where α is a given real-valued closed 2-form in M and β runs over all real-valued 1-forms in M . Any such form $\alpha + d\beta$ is called a representative of γ .) Let

$$\rho R_{11}(M) = H^2_{(1,1)}(M,R) ,$$

(6.22)

$$H^2_{(1,1)}(M,Z) = j^{-1}H^2_{(1,1)}(M,R) .$$

Then we have the theorem:

(B) The image of the homomorphism δ^1 in (6.11) is $H^2_{(1,1)}(M,Z)$.

Since the curvature form $\frac{i}{2\pi}\Omega$ is of type (1,1), we have proved

$$\delta^1 H^1(M,\mathcal{O}^*) \subset H^2_{(1,1)}(M,Z) .$$

To prove inclusion in the other direction, consider the following exact sequence induced by (6.9):

(6.23) $$\ldots \to H^1(M,Z) \xrightarrow{i^1} H^1(M,\mathcal{O}) \xrightarrow{e^1} H^1(M,\mathcal{O}^*) \xrightarrow{\delta^1} H^2(M,Z) \xrightarrow{i^2} H^2(M,\mathcal{O}) \to \ldots .$$

It suffices to prove that $i^2 H^2_{(1,1)}(M,Z) = 0$.

As previously let A^r be the sheaf of germs of complex-valued C^∞ r-forms and C^r be the subsheaf of the germs of A^r closed under d . Also let A^{pq} be the sheaf of germs of C^∞-forms of type (p,q) and C^{pq} be the subsheaf of germs of A^{pq} closed under $\bar\partial$. Thus by definition $C^0 = C$ and $C^{00} = \mathcal{O}$. We have the diagram

$$(6.24)\qquad \begin{array}{ccccccccc} 0 & \longrightarrow & C^r & \xrightarrow{k} & A^r & \xrightarrow{d} & C^{r+1} & \longrightarrow & 0 \\ & & \downarrow \Pi_{0r} & & \downarrow \Pi_{0r} & & \downarrow \Pi_{0,r+1} & & \\ 0 & \longrightarrow & C^{0r} & \xrightarrow{k'} & A^{0r} & \xrightarrow{\bar\partial} & C^{0,r+1} & \longrightarrow & 0 \end{array},$$

where k and k' are inclusions. This diagram is clearly commutative. Moreover both horizontal sequences are exact. The above diagram implies the following commutative diagram of cohomology groups, where the manifold M is omitted in the notation:

$$(6.25)\qquad \begin{array}{ccccc} H^0(C^2)/dH^0(A^1) & \xrightarrow{\Delta^0} & H^1(C^1) & \xrightarrow{\Delta^1} & H^2(C) \\ \downarrow \Pi_{02} & & \downarrow \Pi_{01} & & \downarrow \Pi_{00} \\ H^0(C^{02})/\bar\partial H^0(A^{01}) & \xrightarrow{\tilde\Delta^0} & H^1(C^{01}) & \xrightarrow{\tilde\Delta^1} & H^2(\mathcal{O}) \end{array}$$

Moreover, Δ^0, Δ^1, $\tilde\Delta^0$, $\tilde\Delta^1$ are isomorphisms (cf. §4, in particular (4.11)). We decompose the inclusion i in (6.9) by

$$(6.26)\qquad Z \xrightarrow{h} C \xrightarrow{\Pi_{00}} \mathcal{O}\ ,$$

so that $i = \Pi_{00} \circ h$. For any $\beta \in H^2(Z)$ we have then

$$i^2\beta = \Pi_{00}\, h\beta = \tilde\Delta^1\tilde\Delta^0\, \Pi_{02}\, (\Delta^0)^{-1}(\Delta^1)^{-1}\, h\beta \quad ,$$

by the commutativity of the diagram (6.25). If $\beta \in H^2_{(1,1)}(M,Z)$ we have

$$\Pi_{02}\, (\Delta^0)^{-1}\, (\Delta^1)^{-1}\, h\beta = 0\ ,$$

so that $i^2\beta = 0$. This completes the proof of (B).

To study $H^1(M,\mathcal{O}^*)$ the next step is to consider the subgroup of all $E \in H^1(M,\mathcal{O}^*)$ such that $c(E) = 0$. By the exactness of the sequence (6.23) this is isomorphic to

(6.27) $$H^1(M,\mathcal{O})/i^1H^1(M,Z) \ .$$

For a non-singular projective variety M the group (6.27) is compact and is called the *Picard variety* of M, [19].

The following are some important examples of holomorphic line bundles:

Example 1. The determinant bundle $\Lambda^q(E)$ of a holomorphic vector bundle E of fiber dimension q. If g_{UV} are the transition functions of E, so that g_{UV} are non-singular $(q\times q)$-matrices with elements which are holomorphic functions in $U \cap V$, the bundle $\Lambda^q(E)$ is defined by the transition functions $\det g_{UV}$. If T^* is the cotangent bundle of M and $\dim M = m$, then $\Lambda^m(T^*)$ is called the *canonical bundle* of M; it will be denoted as $K(M)$. If $z_U^1,\ldots,z_U^m$ and $z_V^1,\ldots,z_V^m$ are the local coordinates in U and V respectively, the transition functions of $K(M)$ are the Jacobian determinants

(6.28) $$k_{UV} = \frac{\partial(z_U^1,\ldots,z_U^m)}{\partial(z_V^1,\ldots,z_V^m)} \ .$$

Example 2. Consider the line bundle in Example 2, §1. We will call it the *universal line bundle* over P_m. Here the base space P_m has the covering $\{U_i\}$, and the bundle has the transition functions $g_{ij} = {}_j\zeta^i = \frac{z^i}{z^j}$, $0 \le i,j \le m$, $i \ne j$. The linear form $\Sigma\, a_i z^i$ in $C_{m+1} - 0$, where the a's are constants, has in the local coordinates in $\psi^{-1}(U_i)$ the expression

$$\sum_j a_j z^j = z^i(a_0\, {}_i\zeta^0 + \ldots + \underset{ith}{1} + \ldots + a_m\, {}_i\zeta^m) \ .$$

The expression in parentheses, which is essentially the linear form at the left-hand side in "non-homogeneous" coordinates in U_i, defines a section in the line bundle whose transition functions are

$g'_{ij} = \frac{z^j}{z^i} = ({}_j\zeta^i)^{-1}$. Because of this origin the latter bundle, to be denoted by H , is called the *hyperplane section bundle* of P_m ; it is the negative or dual of the universal line bundle.

Moreover, a holomorphically immersed submanifold $f\colon M \to P_m$ has an induced bundle f^*H , called the *hyperplane section bundle* of M .

§7. Hermitian Geometry and Kählerian Geometry

Let M be a complex manifold of dimension m . M is called *hermitian* if an hermitian structure H is given in its tangent bundle $T(M)$. With the local coordinates $z^1,\ldots,z^m$ a natural frame field is given by

(7.1) $$s_i = \frac{\partial}{\partial z^i}, \qquad 1 \leqq i,j,k,\ell \leqq m ,$$

and this frame is holomorphic. Let

(7.2) $$h_{ik} = H\left(\frac{\partial}{\partial z^i}, \frac{\partial}{\partial z^k}\right) = \bar{h}_{ki} .$$

Then the matrix

(7.3) $$H = {}^t\bar{H} = (h_{ik})$$

is positive definite hermitian.

There are special features arising from the fact that the bundle in question is the tangent bundle. First there is the Kähler form (cf. (2.16))

(7.4) $$\hat{H} = \frac{i}{2}\sum_{j,k} h_{jk}dz^j \wedge d\bar{z}^k ,$$

which is a real-valued form of type (1,1). An hermitian manifold is called *Kählerian* if its Kähler form is closed

(7.5) $$d\hat{H} = 0 .$$

Secondly, let s be a local frame field, holomorphic or not. To s there is uniquely determined a coframe field $\sigma = (\sigma^1,\ldots,\sigma^m)$ such that at every point $x \in M$ the sections $s_1(x),\ldots,s_m(x)$ of s and the sections $\sigma^1(x),\ldots,\sigma^m(x)$ of σ in the cotangent bundle are dual bases. The sections σ^i, being in the cotangent bundle, are complex-valued 1-forms, and they are everywhere linearly independent. Let s' be a new frame field, related to s by (5.20), and let σ' be its dual coframe field. Write

$$
{}^t s = (s_1,\ldots,s_m) , \quad {}^t s' = (s_1',\ldots,s_m') ,
$$

(7.6)

$$
\sigma' = (\sigma'^1,\ldots,\sigma'^m) .
$$

If T_x and T_x^* are respectively the tangent and cotangent spaces at x, we denote their pairing by

(7.7) $$< \xi,\omega > , \quad \xi \in T_x , \quad \omega \in T_x^* .$$

Then we have

(7.8) $$< s_i , \sigma^k > = < s_i' , \sigma'^k > = \delta_i^k .$$

Equation (5.20) can be written

(7.9) $$s_i' = \sum_j g_i^j s_j , \quad g = (g_i^j) .$$

In view of (7.8) we have

(7.10) $$\sigma^i = \sum_j g_j^i \sigma'^j$$

or, in matrix notation,

(7.10a) $$\sigma = \sigma' g .$$

By taking the exterior derivative of (7.10a) and using (5.22), we get

(7.11) $$(d\sigma'-\sigma'\wedge\omega')g = d\sigma - \sigma\wedge\omega .$$

We will call the (1×q)-matrix

(7.12) $$\tau = d\sigma - \sigma\wedge\omega$$

the *torsion matrix*. It is a matrix of complex-valued two forms and follows the transformation law (7.11) under a change of the frame field, holomorphic or not.

(A) Let M be an hermitian manifold. A connection in its tangent bundle is of type (1,0) if and only if its torsion matrix is of type (2,0).

To prove this let σ be the dual coframe field of a holomorphic frame field s. The connection matrix ω relative to s can be written in a unique way as

$$\omega = \omega_1 + \omega_2 ,$$

where ω_1 and ω_2 are matrices of 1-forms of types (1,0) and (0,1) respectively. σ is a matrix of forms of type (1,0) with holomorphic coefficients. The torsion matrix τ in (7.12) is of type (2,0) if and only if

$$\sigma\wedge\omega_2 = 0 .$$

Let

$$\sigma = (\sigma^1,\ldots,\sigma^m) , \quad \omega_2 = (\theta_i^k) , \quad 1 \leqq i,j,k \leqq m .$$

Then the above relation can be written explicitly as

$$\sum_i \sigma^i \wedge \theta_i^k = 0 .$$

Since σ^i are linearly independent, we have

$$\theta_i^k = \sum_j a_{ij}^k \sigma^j ,$$

where a_{ij}^k are functions. But σ^j is of type (1,0), while θ_i^k is of type (0,1) by our hypothesis. It follows that the above relation is equivalent to

$$\theta_i^k = 0 \quad \text{or} \quad \omega_2 = 0 .$$

This proves (A).

The criterion expressed by (A) has the advantage that, unlike the notion of a connection of type (1,0) which is defined in terms of holomorphic frame fields, it has a meaning for C^∞ frame fields. In the study of hermitian manifolds it is desirable to use C^∞ frame fields, for example, the unitary frame fields. It follows from §6 and (A) that an hermitian manifold has a uniquely determined admissible connection in its tangent bundle whose torsion matrix is of type (2,0). When we speak of the connection in an hermitian manifold, this will be the connection meant. It is to be noticed that the curvature matrix of this connection is of type (1,1) (relative to C^∞ frame fields).

(B) An hermitian manifold is Kählerian if and only if the torsion matrix of its connection is zero.

Since both properties are independent of the choice of a frame field, it suffices to verify this theorem by using the natural frame field (7.1). Its dual coframe field is

$$\sigma = (dz^1, \ldots, dz^m) ,$$

so that $d\sigma = 0$. By (6.2) the vanishing of the torsion matrix can be written

$$\sigma \wedge \partial H = 0 ,$$

or, in expanded form,

$$\sum_{i,j} \frac{\partial h_{ik}}{\partial z^j} dz^j \wedge dz^i = 0 , \qquad 1 \leqq i,j,k \leqq m .$$

The latter is equivalent to the conditions

(7.13) $$\frac{\partial h_{ik}}{\partial z^j} - \frac{\partial h_{jk}}{\partial z^i} = 0 .$$

One sees directly that (7.5) and (7.13) are equivalent.

(C) An hermitian manifold is Kählerian if and only if there exists locally a real-valued C^∞-function u, such that its Kähler form can be written

(7.14) $$\hat{H} = i\,\partial\bar{\partial}\,u .$$

It suffices to prove that a Kählerian manifold has the property stated in the theorem, for the form (7.14) is clearly closed. Suppose therefore that $\hat{H}$ is closed. There exists locally a real-valued 1-form ω such that

$$\hat{H} = d\omega .$$

We can write

$$\omega = \alpha + \bar{\alpha} ,$$

where

$$\alpha = \Pi_{1,0}\,\omega , \qquad \bar{\alpha} = \Pi_{0,1}\,\omega .$$

Then

$$d\omega = \partial\alpha + (\partial\bar{\alpha} + \bar{\partial}\alpha) + \bar{\partial}\bar{\alpha} ,$$

where the terms are of types (2,0), (1,1), (0,2) respectively. Since $d\omega$ is of type (1,1), we have

$$\bar{\partial}\bar{\alpha} = 0 .$$

It follows from the Dolbeault-Grothendieck lemma (Theorem (E), §3) that there exists a complex-valued C^∞-function F such that

$$\bar{\alpha} = \bar{\partial}F .$$

Then

$$\hat{H} = d\omega = \partial\bar{\partial}(F-\bar{F}) .$$

The theorem follows by setting $u = -i(F-\bar{F})$.

The most important local properties of an hermitian manifold arise from its curvature matrix. The latter is defined in terms of a frame field. To have the situation under control we list together the formulas giving the effect from a change of the frame field on the various matrices we have introduced (formulas (5.20), (5.23), (5.53), (7.10a)):

$$\begin{aligned} s' &= gs , \\ \Omega' g &= g\Omega , \\ H' &= gH^t\bar{g} , \\ \sigma &= \sigma' g . \end{aligned} \tag{7.15}$$

From the second and third formulas of (7.15) we get

$$\Omega' H' = g\Omega H^t\bar{g} . \tag{7.16}$$

We note that ΩH is skew-hermitian (cf. (5.56)).

Since ΩH is of type (1,1), we set

$$\begin{aligned} \Omega H &= (\Omega_{ik}) \\ \Omega_{ik} &= \sum_{j,\ell} R_{ikj\ell}\sigma^j \wedge \bar{\sigma}^\ell . \end{aligned} \tag{7.17}$$

The skew-hermitian property of ΩH is then expressed by

$$R_{ikj\ell} = \bar{R}_{ki\ell j} . \tag{7.18}$$

Throughout this part of our discussion we suppose as usual that our small Latin indices have the range from 1 to m:

$$1 \leqq i,j,k,\ell,p,q,u,v \leqq m . \tag{7.19}$$

The fourth equation of (7.15) and the equation (7.16) can be written out in detail as follows:

$$\sigma^i = \sum_j g^i_j \sigma'^j .$$

$$\sum_{j,\ell} R'_{ikj\ell}\, \sigma'^j \wedge \bar{\sigma}'^\ell = \sum_{p,q,u,v} g^p_i \bar{g}^q_k\, R_{pquv} \sigma^u \wedge \bar{\sigma}^v ,$$

where the left-hand sides of the second equation are the entries in the matrix $\Omega'H'$. It follows that

$$R'_{ikj\ell} = \sum_{p,q,u,v} g^p_i \bar{g}^q_k g^u_j \bar{g}^v_\ell\, R_{pquv} . \tag{7.20}$$

Let

$$\xi = \sum_i \xi^i s_i = \sum_j \xi'^j s'_j \tag{7.21}$$

be a vector at $x \in M$. The ξ^i and ξ'^j in (7.21) are the components of the vector relative to the frames s and s' respectively. Between them we have the relation

$$\xi^i = \sum_j g^i_j\, \xi'^j \tag{7.22}$$

From (7.20) and (7.22) we get

$$\sum_{i,\ldots,\ell} R'_{ikj\ell}\, \xi'^i\, \bar{\xi}'^k\, \xi'^j\, \bar{\xi}'^\ell = \sum_{i,\ldots,\ell} R_{ikj\ell}\, \xi^i\, \bar{\xi}^k\, \xi^j\, \bar{\xi}^\ell ,$$

so that the common expression is independent of the choice of the frame field. If $\xi \neq 0$, we define

$$R(x,\xi) = 2 \sum_{i,\ldots,\ell} R_{ikj\ell}\, \xi^i\, \bar{\xi}^k\, \xi^j\, \bar{\xi}^\ell \Big/ \left(\sum_{i,k} h_{ik}\, \xi^i\, \bar{\xi}^k\right)^2 \tag{7.23}$$

to be the *holomorphic sectional curvature* at (x,ξ) .

From the second equation of (7.15) we get

$$\mathrm{Tr}\Omega' = \mathrm{Tr}\Omega = \Phi \quad \text{(say).} \tag{7.24}$$

Φ is a form of type (1,1) and is called the *Ricci form* of the hermitian metric. The metric is called *hermitian-einsteinian* if the Ricci form is a multiple of the Kähler form.

Let h^{ik} be the elements of the matrix H^{-1}. By the symmetry relations (7.18) we see that

$$R = \sum_{i,\ldots,\ell} R_{ikj\ell}\, h^{ik}\, h^{j\ell} \tag{7.25}$$

is real; it is also independent of the choice of the frame field. This quantity R is called the *scalar curvature*.

Compact Kählerian manifolds have strong topological restrictions. Perhaps the simplest among them is the following:

(D) The second Betti number of a compact Kählerian manifold is positive.

Corollary. The Hopf and Calabi-Eckmann manifolds $S^{2p+1} \times S^{2q+1}$, $p \geqq 0$, $q \geqq 1$, cannot be given a Kählerian structure. (Cf. §1.)

Since $\hat{H}$ is closed, it determines by de Rham's theorem an element $u \in H^2(M,R)$. To prove (D) we make use of the fact that the 2m-form

$$\hat{H}^m = \hat{H} \wedge \ldots \wedge \hat{H}, \qquad m \text{ times}$$

determines the element $u^m = u \cup \ldots \cup u$ (cup product m times) of $H^{2m}(M,R)$. Using the local expression (7.4), we find

$$\hat{H}^m = \left(\frac{i}{2}\right)^m m!\,(\det H) \bigwedge_j dz^j \wedge d\bar{z}^j . \tag{7.26}$$

Since the matrix H is positive definite, $\det H > 0$. It follows that

$$\int_M \hat{H}^m > 0 ,$$

and $u^m \neq 0$. Therefore $u \neq 0$.

Let M,N be complex manifolds, of dimensions m,n respectively. A continuous mapping $f: M \to N$ is called *holomorphic*, if locally it is defined by expressing the coordinates of the image point as holomorphic functions of those of the original point. f is called an *immersion*, if $m \leqq n$ and if the Jacobian matrix is of rank m everywhere. An immersion f is called an *imbedding*, if it is one-one, i.e., if $f(x) = f(y)$, $x,y \in M$, implies $x = y$. The following is immediate:

(E) Let N be a Kählerian manifold and let $f: M \to N$ be a holomorphic immersion. Then M has a Kählerian structure.

Consider the complex projection space P_n of dimension n . It is known that (cf. §8)

$$\text{(7.27)} \qquad \begin{aligned} &H^{2i}(P_n,Z) \cong Z , &&0 \leqq i \leqq n , \\ &H^k(P_n,Z) = 0 , &&k \text{ odd} . \end{aligned}$$

Moreover we will show in §8 that P_n is Kählerian. In this case, however, there is an additional important fact: The cohomology group $H^2(P_n,R)$ is a real vector space of real dimension 1 and is isomorphic to $jH^2(P_n,Z) \otimes R$, where j is induced by the coefficient homomorphism $j: Z \to R$. In other words, if ζ denotes a generator of $H^2(P_n,Z)$, $j\zeta$ generates $H^2(P_n,R)$ over R . By the multiplication of a constant factor when necessary, we can define on P_n a Kählerian metric such that the cohomology class $u \in H^2(P_n,R)$ determined by the Kähler form belongs to $jH^2(P_n,Z)$. A Kählerian manifold with this property is said to be *of restricted type*.

Under the conditions of Theorem (E) we have the commutative diagram

$$\begin{array}{ccc} H^2(N,Z) & \xrightarrow{j} & H^2(N,R) \\ \downarrow f^* & & \downarrow f^* \\ H^2(M,Z) & \xrightarrow{j} & H^2(M,R) \end{array},$$

where f^* is induced by the mapping f and j is induced by the coefficient homomorphism. Theorem (E) has the following complement:

(E') Let N be a Kählerian manifold of restricted type and let $f: M \to N$ be a holomorphic immersion. Then M is a Kählerian manifold of restricted type.

A theorem of Chow says that a compact complex manifold holomorphically imbedded in P_n is an algebraic variety, i.e., its locus is defined by a finite number of polynomial equations. The imbedding theorem of Kodaira says that a compact Kählerian manifold of restricted type can be holomorphically imbedded in a projective space, [18].

As an example we will study the conditions that the complex torus $\theta = C_m/\Gamma$ (ex. 4, §1) can be given a Kählerian structure of restricted type. Suppose that such a Kählerian structure exists on θ. The latter being a compact connected Lie group, we can integrate the Kählerian metric over θ. The resulting metric will define a Kählerian structure of restricted type which is invariant under the action of θ.

Let $z^1,\ldots,z^m$ be the coordinates in C_m, and let Γ be generated by the vectors

$$(7.28) \qquad \pi_\lambda = (\pi_\lambda^1,\ldots,\pi_\lambda^m), \qquad 1 \leqq \lambda,\mu \leqq 2m,$$

which are linearly independent over R. Let the Kählerian structure of restricted type be given by

$$(7.29) \qquad ds^2 = \sum_{i,k} h_{i\bar{k}} dz^i d\bar{z}^k, \qquad 1 \leqq i,j,k \leqq m,$$

where h_{ik} are C^∞-functions in θ. If the structure is invariant, as we are allowed to assume, h_{ik} are constants. A homology basis for the two-dimensional cycles of θ is formed by the two-dimensional tori $\tau_{\lambda\mu}$, $\lambda < \mu$; $\tau_{\lambda\mu}$ is the quotient of the space spanned by π_λ, π_μ divided by the discrete group generated by π_λ, π_μ. It follows that the metric (7.29) is of restricted type, if and only if

$$(7.30) \qquad g_{\lambda\mu} = -g_{\mu\lambda} = i \sum_{j,k} h_{jk}\left(\pi_\lambda^j \bar{\pi}_\mu^k - \pi_\mu^j \bar{\pi}_\lambda^k\right)$$

are integers. We introduce the matrices

$$(7.31) \qquad G = (g_{\lambda\mu}), \quad H = (h_{ik}), \quad \Pi = (\pi_\lambda^k),$$

so that their orders are (2m×2m), (m×m), and (2m×m) respectively. Then (7.30) can be written

$$(7.30a) \qquad \begin{aligned} G &= \sqrt{-1}(\Pi H\, {}^t\bar{\Pi} - \bar{\Pi}\bar{H}\, {}^t\Pi) \\ &= (\Pi \bar{\Pi}) \begin{pmatrix} \sqrt{-1}H & 0 \\ 0 & -\sqrt{-1}\,\bar{H} \end{pmatrix} \begin{pmatrix} {}^t\bar{\Pi} \\ {}^t\Pi \end{pmatrix} \end{aligned}$$

Taking the inverse matrix of this equation, we get

$$\begin{pmatrix} {}^t\bar{\Pi} \\ {}^t\Pi \end{pmatrix} G^{-1}(\Pi\bar{\Pi}) = \begin{pmatrix} -\sqrt{-1}\,H^{-1} & 0 \\ 0 & \sqrt{-1}\,\bar{H}^{-1} \end{pmatrix}$$

Therefore we have the theorem:

(F) A necessary and sufficient condition for the torus $\theta = C_m/\Gamma$ to have a Kählerian metric of restricted type is that there exists a skew-symmetric matrix G with integral elements such that

$$\text{(7.32)} \qquad \begin{aligned} i\,{}^t\bar{\Pi}\, G^{-1}\, \Pi &> 0 , \\ {}^t\Pi\, G^{-1}\, \Pi &= 0 . \end{aligned}$$

The first condition in (7.32) means that the hermitian matrix at the left-hand side is positive definite.

The conditions (7.32) were first given by Riemann.

We wish to simplify the Riemann conditions (7.32) by proper choices of the basis vectors of C_m and of Γ . Let

$$(z^1,\ldots,z^m) \to (\tilde{z}^1,\ldots,\tilde{z}^m) = (z^1,\ldots,z^m)T$$

be a change of coordinates in C_m , where T is a non-singular $(m\times m)$-matrix with complex elements. Meanwhile, under a change of basis of Γ the matrix Π is transformed according to

$$\Pi \to \tilde{\Pi} = U\Pi ,$$

where U is a unimodular integral matrix. The combined effect of these changes on the matrices is given by

$$\text{(7.33)} \qquad \begin{aligned} \Pi \to \tilde{\Pi} &= U\,\Pi\, T , \\ H \to \tilde{H} &= T^{-1}\, H\, {}^t\bar{T}^{-1} \\ G \to \tilde{G} &= U\, G\, {}^tU . \end{aligned}$$

It is known in the theory of matrices that U can be so chosen that

$$\text{(7.34)} \qquad \tilde{G} = \begin{pmatrix} 0 & D \\ -D & 0 \end{pmatrix}$$

where

$$(7.35) \qquad D = \begin{pmatrix} d_1 & & 0 \\ & \ddots & \\ 0 & & d_m \end{pmatrix}, \qquad d_i \in Z .$$

We can then choose T, so that

$$(7.36) \qquad \tilde{\Pi} = \begin{pmatrix} I \\ \\ \Sigma \end{pmatrix}$$

With $\tilde{G}$, $\tilde{\Pi}$ in place of G, Π, the conditions (7.32) become

$$(7.37) \qquad {}^tZ = Z , \qquad i(\bar{Z}-Z) > 0 ,$$

where

$$(7.38) \qquad Z = \Sigma D .$$

The domain defined by (7.37) is of dimension $m(m+1)/2$; it is called the *Siegel upper half plane* and is known to be biholomorphically equivalent to one of the bounded symmetric domains of Elie Cartan. For $m = 1$ it is the Poincaré half-plane.

An example of a torus not satisfying the Riemann conditions is given in the case $m = 2$ by

$$(7.39) \qquad \Sigma = \begin{pmatrix} \sqrt{-2} & \sqrt{-3} \\ \sqrt{-5} & \sqrt{-7} \end{pmatrix}$$

Then

$$\Sigma D = \begin{pmatrix} \sqrt{-2}d_1 & \sqrt{-5}d_2 \\ \sqrt{-3}d_1 & \sqrt{-7}d_2 \end{pmatrix}, \qquad d_i \in Z$$

For this matrix to be symmetric we must have $d_1 = d_2 = 0$. Then the second condition in (7.37) will not be fulfilled. Thus the corresponding torus will not have a Kählerian structure of restricted type.

In the general case a meromorphic function on the complex torus θ is identical to a 2m-ply periodic meromorphic function in C_m with the period vectors (7.28). All the meromorphic functions on θ form a field. It follows from Kodaira's imbedding theorem that a complex torus satisfying the conditions (7.32) can be holomorphically imbedded in a projective space and is thus by definition an *abelian variety*. At every point x of an abelian variety θ of dimension m there are m meromorphic functions on θ, which are functionally independent at x. On the other hand, there exist complex tori on which every meromorphic function is a constant.

§8. The Grassmann Manifold

Let

$$C_{N+1} = C\times \dots \times C\,, \qquad N + 1 \text{ factors} \tag{8.1}$$

be the complex number space of $N + 1$ dimensions. Let $GL(N+1,C)$ be the general linear group in $N + 1$ complex variables, which we identify with the group of all $(N+1) \times (N+1)$ non-singular matrices with complex elements. Suppose $GL(n+1,C)$ acts on C_{N+1} to the right, as described by

$$(z^0,\dots,z^N) \to (z^0,\dots,z^N)g\,, \quad g \in GL(n+1,C)\,. \tag{8.2}$$

Among the subgroups of $GL(N+1,C)$ are: (1) the unitary group $U(N+1)$, which consists of all matrices g satisfying

$$^t g\bar{g} = I\,, \tag{8.3}$$

where I is the identity matrix; (2) the group $GL(k+1, N-k, C)$, consisting of all non-singular matrices of the form

$$(8.4)\qquad \underbrace{\begin{pmatrix} x & 0 \\ x & x \end{pmatrix}}_{k+1 \quad N-k} \begin{matrix} \} \, k+1 \\ \} \, N-k \end{matrix} ,$$

where the elements at the upper-right corner are zero. The group $GL(k+1, N-k, C)$ is the subgroup of all elements of $GL(N+1, C)$ leaving fixed the $(k+1)$-dimensional subspace of C_{N+1} spanned by the first $k+1$ coordinate vectors.

The space of all $(k+1)$-dimensional linear subspaces of C_{N+1}, $k \geqq 0$, is called a *Grassmann manifold*, to be denoted by $Gr(N,k)$. Using the projection

$$(8.5)\qquad \psi\colon C_{N+1} - 0 \to P_N$$

in Example 2, §1, the notation suggests that it is also the space of all k-dimensional linear (projective) subspaces in P_N.

From the above discussion $Gr(N,k)$ can be represented as a right coset space in two different ways:

$$(8.6)\qquad Gr(N,k) = \frac{GL(N+1,C)}{GL(k+1,N-k,C)} = \frac{U(N+1)}{U(k+1)\times U(N-k)} .$$

The first representation shows that it is a complex manifold of dimension $(k+1)(N-k)$. The second representation shows that it is compact.

An element of $Gr(N,k)$ can be given by a non-zero decomposable $(k+1)$-vector

$$(8.7)\qquad \Lambda = X_0 \wedge X_1 \wedge \ldots \wedge X_k \neq 0$$

defined up to a constant factor. If $e_0^0, \ldots, e_N^0$ denote a fixed frame in C_{N+1}, we can write

$$(8.8)\qquad \Lambda = \sum_{\alpha} p_{\alpha_0 \ldots \alpha_k}\, e_{\alpha_0}^0 \wedge \cdots \wedge e_{\alpha_k}^0 , \qquad 0 \leqq \alpha_0, \ldots, \alpha_k \leqq N ,$$

where the p's are skew-symmetric in their indices. The $p_{\alpha_0 \ldots \alpha_k}$ are called the *Cayley-Plücker-Grassmann* coordinates in $Gr(N,k)$. By considering $p_{\alpha_0 \ldots \alpha_k}$ as the homogeneous coordinates of a projective space of dimension $\binom{N+1}{k+1} - 1$, we get an imbedding of $Gr(N,k)$ in the latter.

We propose to study the topological properties of $Gr(N,k)$. Our main step is to obtain a cell decomposition of $Gr(N,k)$ by means of the *Schubert varieties*. This was first accomplished by C. Ehresmann in 1934. Let

$$0 \leqq a_0 \leqq a_1 \leqq \ldots \leqq a_k \leqq N - k \tag{8.9}$$

by a sequence of integers, and let

$$L_{a_0} \subset L_{a_1+1} \subset \ldots \subset L_{a_k+k} \subset P_N \tag{8.10}$$

be a nested sequence of linear spaces whose dimensions are given by the subscripts. (We will deal with linear spaces of P_N; their images under ψ^{-1} will have one dimension higher.) A *Schubert variety* $(a_0 a_1 \ldots a_k)$ is the set of all k-dimensional linear spaces $X \in Gr(N,k)$ such that

$$\dim(X \cap L_{a_j+j}) \geqq j, \qquad 0 \leqq j \leqq k . \tag{8.11}$$

By definition, $(a_0 a_1 \ldots a_k)$ is a closed subset of $Gr(N,k)$ and is determined by the a's up to a projective collineation. Its (complex) dimension is

$$\dim(a_0 a_1 \ldots a_k) = a_0 + a_1 + \ldots + a_k . \tag{8.12}$$

<u>Examples</u>: 1. $(N-k \ldots N-k) = Gr(N,k)$;

2. $(0 \ldots 0) = L_k$;

3. $(0 \ldots 01 \ldots 1)$ (r ones) is the set of all X satisfying the condition

(8.13) $$L_{k-r} \subset X \subset L_{k+1} ,$$

where L_{k-r} and L_{k+1} are fixed linear spaces of dimensions $k-r$ and $k+1$ respectively.

We take a fixed sequence of linear spaces in P_N :

(8.14) $$L_0 \subset L_1 \subset \ldots \subset L_{N-1} \subset P_N$$

and suppose the Schubert varieties are constructed from the linear spaces of this sequence. Put

(8.15)

$$(a_0 \ldots a_k)^* = (a_0 \ldots a_k) - \sum_{a_{j-1}<a_j} (a_0 \ldots a_{j-1} a_j - 1 \ldots a_k) ,$$

$$(a_{-1} = 0) .$$

Then any $X \in Gr(N,k)$ belongs to a unique $(a_0 \ldots a_k)^*$. That is, the sets $(a_0 \ldots a_k)^*$ are mutually disjoint and their union is $Gr(N,k)$.

(A) $(a_0 \ldots a_k)^*$ is an open cell of real dimension $2(a_0 + \ldots + a_k)$.

<u>Example</u>. For $k = 0$ we have

(8.16) $$P_N = (N)^* + (N-1)^* + \ldots + (1)^* + (0)^*$$

$$= (P_N - L_{N-1}) + (L_{N-1} - L_{N-2}) + \ldots + (L_1 - L_0) + L_0 .$$

That is, P_N is a union of cells of dimensions $0,2,4,\ldots,2N$ respectively, which are mutually disjoint.

We prove (A) by induction on k. The example shows that it is true for $k = 0$. For definiteness we suppose $a_0 > 0$; the treatment of the case $a_0 = 0$ requires only slight modifications. We take a hyperplane π in P_N such that

$$\pi \cap L_{a_0} = L_{a_0-1} \quad ,$$

$$\pi \cap L_q = L'_{q-1} \; , \qquad a_0 < q \; ,$$

where L'_{q-1} is of dimension $q-1$. Consider the set

$$\Sigma = (a_0 a_1 \ldots a_k) - (a_0-1 \; a_1 \ldots a_k)$$

If $X \in \Sigma$, it meets L_{a_0}, but not L_{a_0-1}. Hence it meets $L_{a_0} - L_{a_0-1}$ in exactly one point y (say). Moreover the intersection

$\xi = X \cap \pi$ is of dimension $k-1$, satisfying $\xi \cap L_{a_0-1} = \emptyset$. We therefore have the continuous mapping

$$\phi: \Sigma \to (L_{a_0} - L_{a_0-1}) \times A \tag{8.17}$$

defined by

$$\phi(X) = (y,\xi) \; ,$$

where A is the subset of the Grassmann manifold $Gr(N-1,k-1)$ of all $(k-1)$-dimensional linear spaces in π such that $\xi \cap L_{a_0-1} = \emptyset$.

To describe the image $\phi(\Sigma)$ we consider in π the sequence of linear spaces

$$L_{a_0-1} \subset L'_{a_0} \subset \ldots \subset L'_{N-2} \subset \pi \; . \tag{8.18}$$

The Schubert varieties of $Gr(N-1, k-1)$ to be considered will be defined relative to the sequence (8.18) and will be denoted by the same symbols with dashes. We have, for $j \geqq 1$,

$$\dim(\xi \cap L'_{a_j+j-1}) = \dim(X \cap \pi \cap L_{a_j+j}) \geqq j-1 ,$$

so that $\xi \in (a_1 \ldots a_k)'$. Conversely, if

$$(y,\xi) \in (L_{a_0} - L_{a_0-1}) \times (a_1 \ldots a_k)' ,$$

they span a k-dimensional space X whose intersection

$$X \cap L_{a_j+j} \qquad (j \geqq 1)$$

contains y and $\xi \cap L'_{a_j+j-1}$ and is hence of dimension $\geqq j$. If moreover, $\xi \in A$, then $X \in \Sigma$. Thus ϕ is a homeomorphism of Σ onto the set

$$(L_{a_0} - L_{a_0-1}) \times \{(a_1 \ldots a_k)' \cap A\} .$$

Analogous to (8.15) we set

$$(a_1 \ldots a_k)'^* = (a_1 \ldots a_k)' - \sum_{a_{j-1}<a_j} (a_1 \ldots a_{j-1} a_j - 1 \ldots a_k)' .$$

Then $\xi \in (a_1 \ldots a_k)'^*$ implies that $\xi \cap L'_{a_1-1} = \emptyset$ so that $\xi \in A$. Our induction hypothesis says that $(a_1 \ldots a_k)'^*$ is an open cell.

The homeomorphism ϕ depends only on a_0 and on the choice of π ; it is independent of the integers $a_1, \ldots, a_k$, provided that the conditions (8.9) are fulfilled. It follows that, for $j \geqq 1$, $a_{j-1} < a_j$, ϕ restricts to a homeomorphism

$$\begin{array}{c} \phi\colon (a_0 a_1 \ldots a_{j-1} a_j - 1 \ldots a_k) - (a_0 - 1\ a_1 \ldots a_{j-1} a_j - 1 \ldots a_k) \\ \big\downarrow \\ (L_{a_0} - L_{a_0-1}) \times \{(a_1 \ldots a_{j-1} a_j - 1 \ldots a_k)' \cap A\} . \end{array}$$

From this we see easily that ϕ establishes a homeomorphism between $(a_0 \ldots a_k)^*$ and $(L_{a_0} - L_{a_0-1}) \times (a_1 \ldots a_k)'^*$. This proves (A).

The Schubert varieties relative to the sequence (8.14) give a cell decomposition of $Gr(N,k)$ whose cells are all of even dimensions. From known theorems in algebraic topology (cf. [9]) we are thus able to draw the following conclusions on the topological properties of $Gr(N,k)$:

(B) The Schubert varieties are cycles. $Gr(N,k)$ is simply connected. It has no torsion coefficients and its homology groups of odd dimensions are zero. A homology basis of $Gr(N,k)$ of dimension $2r$ is formed by the Schubert varieties $(a_0 a_1 \ldots a_k)$, where a_0 , $a_1, \ldots, a_k$ run over all sets of integers satisfying $0 \leqq a_0 \leqq a_1 \leqq \ldots \leqq a_k \leqq N - k$ and $a_0 + a_1 + \ldots + a_k = r$.

Example. $Gr(3,1)$ is the space of all lines in P_3 . Its (complex) dimension is 4. Its Schubert cycles of different dimensions are respectively

$$(00), (01), (11), (02), (12), (22) .$$

Hence its Betti numbers are

$$b^0 = b^8 = 1, \quad b^2 = b^6 = 1, \quad b^4 = 2,$$

where the superscripts indicate the (real) dimensions.

Of geometrical significance is the structure of the homology or cohomology rings of $Gr(N,k)$, i.e., the intersection properties of the Schubert varieties. These are at the basis of enumerative geometry and have been completely determined (cf. [8], [23]). We will, however, not enter into this question.

The Grassmann manifold has been playing an important rôle in recent developments of mathematics, because it is a so-called *classifying space* for complex vector bundles. In fact, when $Gr(N,k)$ is considered to be the manifold of all (k+1)-dimensional linear subspaces through the origin of C_{N+1} , it is the base space of a complex vector bundle whose fibers are these linear subspaces themselves.

More precisely, let $\Lambda \in Gr(N,k)$ as defined by (8.7) and let $v \in C_{n+1}$ be such that $v \wedge \Lambda = 0$, i.e., $v \in \Lambda$. Also let

$$E_0 = \{(v,\Lambda) | v \wedge \Lambda = 0\} . \tag{8.19}$$

Then

$$\psi_0 : E_0 \to Gr(N,k) , \tag{8.20}$$

with the projection ψ_0 defined by

$$\psi_0(v,\Lambda) = \Lambda , \tag{8.21}$$

is a complex vector bundle with the fiber dimension $k + 1$.

The bundle (8.20) has an important property. To describe it we define a *(k+1)-frame* to be an ordered set of $k + 1$ vectors $e_0, \ldots, e_k \in C_{N+1}$ such that $e_0 \wedge \ldots \wedge e_k \neq 0$. The space of all (k+1)-frames in C_{N+1} is called a *Stiefel manifold,* to be denoted by $St(N+1, k+1)$. It is the total space of a fiber bundle

$$\lambda : St(N+1, k+1) \to E_0 \tag{8.22}$$

over E_0 , with the projection λ defined by

$$\lambda(e_0,\ldots,e_k) = (e_0, e_0 \wedge \ldots \wedge e_k) \in E_0 . \tag{8.23}$$

The total space $St(N+1, k+1)$ has a string of vanishing homotopy groups expressed by

$$\pi_i(St(N+1, k+1)) = 0 , \qquad i \leqq 2N - 2k \tag{8.24}$$

(cf. [11], p. 134). Following Steenrod's terminology the bundle (8.20) is (2N-2k+1)-universal in the following sense: Let M be a compact manifold of real dimension $\leqq 2N - 2k$. The equivalence classes of complex vector bundles of fiber dimension $k+1$ over M are in one-one correspondence with the homotopy classes of continuous mappings $f: M \to Gr(N,k)$, the correspondence being established by assigning to each mapping f the bundle f^*E_0 induced from E_0 .

On account of this theorem the bundle (8.20) is called a *universal bundle* and its base space a *classifying space*.

The *universal rth Chern class* $\tilde{c}_r$, $0 \leq r \leq k+1$, is the element of $H^{2r}(Gr(N,k),Z)$ such that its value is 1 over the Schubert cycle (0 ...01 ... 1) (r ones) and is 0 over all other Schubert cycles. By the above theorem if $\psi: E \to M$ is a complex vector bundle with fiber dimension $k+1$, it is induced from E_0 by a mapping $f: M \to Gr(N,k)$ (real dim $M \leq 2N - 2k$) , and f is defined up to a homotopy. It follows that $f^*\tilde{c}_r \in H^{2r}(M,Z)$ is completely determined by the bundle E . We define .

(8.25) $$c_r(E) = f^*\tilde{c}_r \in H^{2r}(M,Z) , \quad 0 \leq r \leq k + 1 ;$$

$c_r(E)$ is called the *rth Chern class of* E. Clearly $c_0(E) = 1$.

In applications it will be essential to identify $c_r(E)$ with geometric or analytic invariants defined in other ways. We will sketch one such application without insisting on details. Let M be a compact almost complex manifold of real dimension $2n$. Its tangent bundle $T(M)$ is then a complex vector bundle over M with fiber dimension n . Then we have

(8.26) $$c_n(T(M))\cdot M = \chi(M) ,$$

where the left-hand side stands for the value of $c_n(T(M))$ on the fundamental cycle of M and the right-hand side $\chi(M)$ is the Euler-Poincaré characteristic of M .

To see this we consider the universal Chern class $\tilde{c}_n \in H^{2n}(Gr(N,n-1),Z)$, with N sufficiently large. By definition this is the class which has the value one over the Schubert cycle (1 ... 1) (n ones) and the value zero over all other Schubert cycles. By Poincaré duality this can be realized by taking a fixed Schubert cycle (N-n, ..., N-n) of complementary real dimension $2n(N - n)$ and taking its intersection with the Schubert cycles of real dimension $2n$. By definition $(N-n \ldots N-n)$ consists of all the n-dimensional linear spaces through 0 in C_{N+1} which lie in a fixed hyperplane L of dimension N. Let v_0 be a vector through 0 in C_{N+1} orthogonal to L. By using the mapping $f\colon M \to Gr(N,n-1)$ and by taking the orthogonal projection of v_0 to $f(x)$, $x \in M$, we define a vector field over M, which will have singularities exactly at the points $x \in M$ such that $f(x) \in L$. One verifies that $c_n(T(M)) \cdot M$ is equal to the sum of the indices at the singularities of a vector field with a finite number of singularities. This proves (8.26).

By studying the homotopy groups of the unitary group, Bott proved the theorem: Let E be a complex vector bundle of fiber dimension n over the 2n-sphere S^{2n}. Then $c_n(E) \cdot S^{2n}$ is divisible by $(n-1)!$

If S^{2n} has an almost complex structure and E is the tangent bundle, then by (8.26) $c_n(T(S^{2n})) \cdot S^{2n}$ is equal to 2, the Euler-Poincaré characteristic of S^{2n}. It follows from Bott's theorem that S^{2n} has an almost complex structure only when $n \leqq 3$. On the other hand, it can be proved by a different method that S^4 does not have an almost complex structure. Thus S^2 and S^6 are the only even-dimensional spheres which have almost complex structures.

We now study the geometry in $Gr(N,k)$. For this purpose it is necessary to introduce an hermitian structure in the bundle (8.20). This is most easily achieved by introducing in C_{N+1} the hermitian scalar product

$$(8.27) \qquad (Z,W) = \overline{(W,Z)} = z_0\bar{w}_0 + \ldots + z_N\bar{w}_N ,$$

where

$$Z = (z_0,\ldots,z_N) \in C_{N+1} , \qquad W = (w_0,\ldots,w_N) \in C_{N+1}$$

This induces an hermitian structure in E_0 in an obvious way.

The definition (8.27) can be extended to decomposable (k+1)-vectors. In fact, let

$$(8.29) \qquad \Lambda = X_0 \wedge \ldots \wedge X_k , \qquad M = Y_0 \wedge \ldots \wedge Y_k .$$

We define the hermitian scalar product

$$(8.30) \qquad (\Lambda,M) = \det(X_\alpha, Y_\beta) , \qquad 0 \leqq \alpha,\beta \leqq k .$$

The product (Λ,M) in (8.30) depends only on the (k+1)-vectors and is independent of the ways that they are decomposed in (8.29). By the hermitian property of (Λ,M) and the fact that any (k+1)-vector is a linear combination of decomposable (k+1)-vectors, the definition of (Λ,M) is extended to arbitrary (k+1)-vectors Λ,M . For simplicity of writing we will introduce the notations

$$(8.31) \qquad |\Lambda,M| = |(\Lambda,M)| , \qquad |\Lambda| = +(\Lambda,\Lambda)^{1/2} .$$

$|\Lambda|$ is called the *norm* of Λ . Clearly the norm determines the scalar product. The quotient $|\Lambda,M|/|\Lambda||M|$ depends only on the elements of $Gr(N,k)$ determined by the (k+1)-vectors Λ, M, and we have the Schwarz inequality

$$(8.32) \qquad |\Lambda,M| \leqq |\Lambda||M| .$$

Utilizing the scalar product (8.27) we will restrict ourselves to unitary frames. A *unitary (h+1)-frame* is an ordered set of $h + 1$ vectors $Z_0, Z_1, \ldots, Z_h$ satisfying

$$(8.33) \qquad (Z_i,Z_j) = \delta_{ij} , \qquad 0 \leqq i,j \leqq h .$$

If $h = N$, we will call it simply a *unitary frame*. We will identify $U(N+1)$ with the space of all unitary frames. Then we have the fiberings

$$(8.34) \qquad U(N+1) \xrightarrow{\lambda} St(N+1, k+1) \xrightarrow{\mu} Gr(N,k) ,$$

where $St(N+1, k+1)$ is the Stiefel manifold of all unitary $(k+1)$-frames in C_{N+1} and the projections λ, μ are defined by

$$\lambda(Z_0, Z_1, \ldots, Z_N) = (Z_0, Z_1, \ldots, Z_k) \, ,$$

(8.35)

$$\mu(Z_0, Z_1, \ldots, Z_k) = Z_0 \wedge Z_1 \wedge \ldots \wedge Z_k \, ,$$

the last $(k+1)$-vector defining an element of $Gr(N,k)$.

In $U(N+1)$ we put

(8.36) $$\theta_{AB} = (dZ_A, Z_B) \, , \qquad 0 \leqq A,B,C \leqq N \, .$$

From the orthogonality relations (8.33) we get by differentiation

(8.37) $$\theta_{AB} + \bar{\theta}_{BA} = 0 \, .$$

Equation (8.36) can also be written

(8.38) $$dZ_A = \sum_B \theta_{AB} Z_B \, .$$

Taking its exterior derivative, we get

(8.39) $$d\theta_{AB} = \sum_C \theta_{AC} \wedge \theta_{CB} \, .$$

These are called the *Maurer-Cartan equations* of the unitary group $U(N+1)$.

Under the projection $\mu \circ \lambda$ in (8.34) the differential forms of $Gr(N,k)$ are mapped into forms of $U(N+1)$, and this mapping is an isomorphism, i.e., a form ω on $Gr(N,k)$ is completely determined by its image $(\mu \circ \lambda)^* \omega$. We will utilize this fact by studying the forms on $U(N+1)$ and consider a relation to be on $Gr(N,k)$ when all the forms involved belong to the image of $(\mu \circ \lambda)^*$. Moreover for simplicity the mapping $(\mu \circ \lambda)^*$ will be omitted in the formulas.

With these conventions, let Λ be a decomposable $(k+1)$-vector and let

(8.40) $$\Lambda_0 = \Lambda / |\Lambda| \, ,$$

so that Λ_0 is a unit (k+1)-vector. We write

(8.41) $$\Lambda_0 = Z_0 \wedge \ldots \wedge Z_k ,$$

$Z_0,\ldots,Z_k$ being a unitary (k+1)-frame. Then we get, by means of (8.38),

(8.42) $$(d\Lambda_0,\Lambda_0) = \sum_\alpha \theta_{\alpha\alpha} = -\sum_\alpha \bar{\theta}_{\alpha\alpha}$$

$$(d\Lambda_0,d\Lambda_0) = +\left(\sum_\alpha \theta_{\alpha\alpha}\right)\left(\sum_\alpha \bar{\theta}_{\alpha\alpha}\right) + \sum_{\alpha,r} \theta_{\alpha r}\bar{\theta}_{\alpha r} ,$$

$$0 \leqq \alpha \leqq k, \quad k+1 \leqq r \leqq N ,$$

where the multiplication of differential forms is in the sense of ordinary commutative multiplication. It follows that

$$(d\Lambda_0,d\Lambda_0) - (d\Lambda_0,\Lambda_0)(\Lambda_0,d\Lambda_0) = \sum_{\alpha,r} \theta_{\alpha r}\bar{\theta}_{\alpha r}$$

By substituting the expression in (8.40), we get

(8.43) $$\frac{1}{|\Lambda|^4}\{(\Lambda,\Lambda)(d\Lambda,d\Lambda) - (d\Lambda,\Lambda)(\Lambda,d\Lambda)\} = \sum_{\alpha,r} \theta_{\alpha r}\bar{\theta}_{\alpha r} .$$

This defines an hermitian structure in Gr(N,k) . In fact, the left-hand side of (8.43) shows that it is hermitian and the right-hand side shows that the metric is positive definite.

The Kähler form of (8.43) is

$$\hat{H}_k = \frac{i}{2}\sum_{\alpha,r} \theta_{\alpha r}\wedge\bar{\theta}_{\alpha r} = \frac{1}{2i}\sum_{\alpha,r} \theta_{\alpha r}\wedge\theta_{r\alpha} = \frac{1}{2i}\, d\left(\sum_\alpha \theta_{\alpha\alpha}\right) .$$

It is therefore closed, and the metric (8.43) is Kählerian. By (8.40) and (8.42) we can further write

(8.45) $$\sum_\alpha \theta_{\alpha\alpha} = (\partial-\bar{\partial}) \log |\Lambda| ,$$

so that

(8.46) $$\hat{H}_k = i\, \partial\bar{\partial} \log |\Lambda| .$$

We summarize the results in the theorem:

(C) The Grassmann manifold $Gr(N,k)$ has a Kählerian structure invariant under the action of $U(N+1)$. Its Kähler form is equal to π times the curvature form of the hyperplane section bundle over $Gr(N,k)$ defined by the imbedding by the Cayley-Plücker-Grassmann coordinates and the hermitian norm $|\Lambda|$.

The first statement has been proved. The second statement may need some explanation. All the $(k+1)$-vectors Λ of C_{N+1}, decomposable or not, form a complex vector space C_ν of dimension $\nu = \binom{N+1}{k+1}$. As in Example 2, §1 and Example 2, §6, $C_\nu - \{0\} \to P_{\nu-1}$ defines the universal line bundle over $P_{\nu-1}$ and an hermitian structure is introduced in this bundle by the norm $|\Lambda|$. The restriction of this bundle to $Gr(N,k) \subset P_{\nu-1}$ is the negative of the hyperplane section bundle meant in the theorem, and $|\Lambda|^{-1}$ defines an hermitian structure on it.

By (6.5) the curvature form of this bundle is $\frac{1}{2\pi i}\partial\bar{\partial}\log h$, where $h = |\Lambda|^{-2}$ is the square of the norm of a local holomorphic section. It is therefore equal to $\frac{i}{\pi}\partial\bar{\partial}\log|\Lambda| = +\frac{1}{\pi}\hat{H}_k$, by (8.46).

This proves the second statement in (C).

Remark. Consider the universal bundle (8.20). Let V be a neighborhood in $Gr(N,k)$ and let Z_A, $0 \leq A \leq N$, be a frame field over V into $U(N+1)$. Then $Z_0, \ldots, Z_k$ define a frame field of the bundle E_0 over V. The matrix $(\theta_{\alpha\beta})$, $0 \leq \alpha, \beta \leq k$, depends only on the $(k+1)$-frame field Z_α and follows the transformation law (5.22) under a change of the frame field. It therefore defines a connection in the bundle E_0. The curvature matrix of this connection is $\Theta = (\Theta_{\alpha\beta})$, where, by (8.39),

(8.47)

$$\Theta_{\alpha\beta} = d\theta_{\alpha\beta} - \sum_{\gamma} \theta_{\alpha\gamma} \wedge \theta_{\gamma\beta} = \sum \theta_{\alpha r} \wedge \theta_{r\beta} = - \sum \theta_{\alpha r} \wedge \bar{\theta}_{\beta r} ,$$

$$0 \leqq \alpha,\beta,\gamma \leqq k , \quad k + 1 \leqq r \leqq N .$$

It follows that

$$\Theta + {}^t\bar{\Theta} = 0$$

and hence, as in (5.59), that the determinant $\det(I + \frac{i}{2\pi}\Theta)$ is real. Using the notation of §5, we have therefore $(\mathrm{Im}P_r)(\Theta) = 0$, $0 \leqq r \leqq k + 1$.

By actual integration one can show that

(8.48) $$\binom{k+1}{r} \int_{(0\ldots01\ldots1)} (\mathrm{Re}\; P_r)(\Theta) = 1 ,$$

where $(0\ldots01\ldots1)$ is the Schubert cycle with r ones and that the same integral over any other Schubert cycle is zero. This means that the element of $H^{2r}(Gr(N,k),R)$ defined by $\binom{k+1}{r}$ $(\mathrm{Re}\; P_r)(\Theta)$ via the de Rham isomorphism is $j\tilde{c}_r$, where $\tilde{c}_r$ is the rth universal Chern class and

$$j\colon H^{2r}(Gr(N,k),Z) \to H^{2r}(Gr(N,k),R)$$

is induced by the coefficient homomorphism.

Let M be a compact manifold and $\psi\colon E \to M$ be a complex vector bundle of fiber dimension k+1 induced from E_0 by the mapping $f\colon M \to Gr(N,k)$. Then the above relationship remains true for the induced connection. By Theorem (B), §5, we conclude that

$$\binom{k+1}{r} P_r(\Omega) , \quad 0 \leqq r \leqq k + 1 ,$$

where Ω is the curvature matrix of any connection in E, corresponds to the Chern class $jc_r(E)$ by the de Rham isomorphism. This is a relationship between the curvature of a connection of a complex vector bundle and its characteristic classes and contains as a special case the Gauss-Bonnet formula in high dimensions.

§9. Curves in a Grassmann Manifold

As in §8 we will denote by $Gr(N,k)$ the Grassmann manifold of all k-dimensional linear subspaces of the projective space P_N of dimension N. Let M be a one-dimensional complex manifold or Riemann surface. A *holomorphic curve* in $Gr(N,k)$ is a holomorphic mapping $f: M \to Gr(N,k)$. In particular, a holomorphic curve $f: M \to Gr(1,0) = P_1$ is a meromorphic function on M, given by the ratio of the homogeneous coordinates in P_1.

Let Λ be a non-zero decomposable (k+1)-vector which defines an element of $Gr(N,k)$, so that Λ is determined up to a factor. Let Σ_B be the subset of all $\Lambda \in Gr(N,k)$ such that

$$(\Lambda, B) = 0, \tag{9.1}$$

where B is a fixed (k+1)-vector, decomposable or not. Σ_B is then a submanifold of codimension one in $Gr(N,k)$.

This set Σ_B has a simple geometrical interpretation if B is decomposable. In fact, we shall say that two points $Z, W \in P_N$ are *orthogonal* if $(Z,W) = 0$. (We will identify points of P_N with their homogeneous coordinate vectors, and the same with elements of $Gr(N,k)$.) If B is decomposable, it determines an element $B \in Gr(N,k)$. Associated to the latter there is an element $B^{\perp} \in Gr(N,N-k-1)$, which is *completely orthogonal* to B, i.e., $B^{\perp}$ is the set of all points of P_N which are orthogonal to all points of B. This relationship is obviously symmetrical: $(B^{\perp})^{\perp} = B$. For a decomposable B, Σ_B is the set of all $\Lambda \in Gr(N,k)$ which meet $B^{\perp}$.

Since (Λ,B) is holomorphic in $Gr(N,k)$ we have

$$\partial\bar{\partial} \log|\Lambda,B|^2 = \partial\bar{\partial}\{\log(\Lambda,B) + \log(B,\Lambda)\} = 0 .$$

Combining with (8.46) we can write

$$\hat{H}_k = i\,\partial\bar{\partial}\log\frac{|\Lambda|\cdot|B|}{|\Lambda,B|} .$$

This formula is valid in $Gr(N,k) - \Sigma_B$ (while (8.46) is valid only locally).

Let

(9.3) $$f: M \to Gr(N,k)$$

be a holomorphic curve such that the image $f(M)$ does not belong to Σ_B. Let D be a compact domain in M with smooth boundary set ∂D and consider the restriction $f|D$. Since the points of the set $f^{-1}(f(D) \cap \Sigma_B)$ are the zeros of the holomorphic function $(\Lambda(\zeta),B)$, $\zeta \in D$, which is itself not identically zero, the set $f^{-1}(f(D) \cap \Sigma_B)$ consists only of a finite number of points. Let $\zeta_0 \in D$ be such a point, with ζ the local coordinate in a neighborhood of ζ_0. We define the *order of* ζ_0 to be the order of zero of the holomorphic function

$$w(\zeta) = (\Lambda(\zeta),B)$$

at ζ_0. The sum of the orders of all such points we define to be the order $n(D,B)$ of the domain D relative to B or the intersection number of $f(D)$ and Σ_B. About the point ζ_0 draw a circle γ_ε of radius ε. By a local consideration it is easily seen that the order of ζ_0 is equal to

$$\frac{1}{2\pi i}\lim_{\varepsilon\to 0}\int_{\gamma_\varepsilon}(\partial-\bar{\partial})\log|w| .$$

Define

(9.4) $$v(D) = \frac{1}{\pi}\int_{f(D)}\hat{H}_k ,$$

so that $v(D)$ is the normalized volume of $f(D)$. By applying

Stokes' theorem to the formula (9.2), we get the theorem:

(A) (First main theorem for holomorphic curves.) Let $f\colon M \to Gr(N,k)$ be a holomorphic curve and let D be a compact domain of M with smooth boundary ∂D. Let B be a fixed $(k+1)$-vector such that $f(\partial D) \cap \Sigma_B = \emptyset$. Then

$$n(D,B) - v(D) = \frac{1}{2\pi i} \int_{f(\partial D)} (\partial-\bar{\partial}) \log \frac{|\Lambda, B|}{|\Lambda|\cdot|B|},$$

where $n(D,B)$ is the order of D relative to B and $v(D)$ is the normalized volume of $f(D)$.

Corollary. If M is a compact holomorphic curve without boundary in $Gr(N,k)$, then

$$n(M,B) = v(M), \tag{9.6}$$

i.e., $f(M)$ meets all Σ_B the same number $v(M)$ of times.

With the classical theory of meromorphic functions as an example, we will study holomorphic curves (9.3) where M is non-compact. The right-hand side of (9.5) motivates us to introduce the operator

$$d^c = i(\bar{\partial}-\partial). \tag{9.7}$$

d^c acts on complex-valued C^∞-forms and is of degree 1, i.e., it maps a form of degree r to a form of degree $r+1$. It is also a real operator in the sense that it maps a real form into a real form. We have

$$dd^c = 2i\partial\bar{\partial}. \tag{9.8}$$

A smooth function τ on M is called *harmonic*, if

$$dd^c\tau = 0. \tag{9.9}$$

If τ and g are two smooth functions on M, we have

$$d\tau \wedge d^C g - dg \wedge d^C\tau = 2i(\partial g \wedge \partial\tau - \bar{\partial} g \wedge \bar{\partial}\tau) .$$

Since M is one-dimensional, the right-hand side is zero and we have

(9.10) $$d\tau \wedge d^C g = dg \wedge d^C \tau .$$

Example. Suppose $M = C^* = C - \{0\}$. Let $\zeta \in C^*$, $\zeta = re^{i\theta}$, $r \neq 0$. Then

$$\log \zeta = \log r + i\theta ,$$

$$\log \bar{\zeta} = \log r - i\theta .$$

If $\tau = \log r$, we have

$$d\tau = d \log r = 1/2\{d \log \zeta + d \log \bar{\zeta}\}$$

and

(9.11) $$d^C\tau = - 1/2\{d \log \zeta - d \log \bar{\zeta}\} = d\theta .$$

Now suppose τ is a real-valued smooth function on M. With a local coordinate ζ we have

(9.12) $$d\tau = \tau_\zeta \, d\zeta + \tau_{\bar{\zeta}} \, d\bar{\zeta} ,$$

so that

(9.13) $$d^C\tau = i(-\tau_\zeta d\zeta + \tau_{\bar{\zeta}} d\bar{\zeta})$$

and

(9.14) $$d\tau \wedge d^C\tau = 2\tau_\zeta \, \tau_{\bar{\zeta}}(i d\zeta \wedge d\bar{\zeta}) .$$

An *exhaustion function* on M is a smooth function $\tau: M \to R^+$ (= set of real numbers $\geqq 0$) which satisfies the conditions: (1) the mapping τ is proper, i.e., $\tau^{-1}(A)$ is compact whenever $A \subset R^+$ is compact; (2) the critical points of τ, i.e., points at which $d\tau = 0$, are isolated. An example of a function satisfying the second condition is a real-valued harmonic function.

Suppose that M has an exhaustion function τ ; such a function exists on every non-compact manifold, cf. [25], p. 36. Let

$$D_u = \{\zeta \in M \mid \tau(\zeta) \leqq u\} \tag{9.15}$$

and write

$$n(D_u, B) = n(u,B), \qquad v(D_u) = v(u). \tag{9.16}$$

Then (9.5) can be written

$$n(u,B) - v(u) = \frac{1}{2\pi}\int_{f(\partial D_u)} d^c \log \frac{|\Lambda, B|}{|\Lambda|\cdot|B|}. \tag{9.17}$$

By (9.10) we have

$$d^c g \equiv \frac{\partial g}{\partial \tau} d^c\tau, \quad \text{mod } d\tau.$$

The boundary ∂D_u is the curve $\tau = u$, along which $d\tau = 0$. Hence the right-hand side of (9.17) is equal to

$$\frac{1}{2\pi}\int_{\partial D_u} f^* \frac{\partial}{\partial\tau} \log \frac{|\Lambda, B|}{|\Lambda|\cdot|B|} \; d^c\tau.$$

By an argument which we will not give here (cf. [13], p. 70 or [15]), the integral and the differential operator $\partial/\partial\tau$ can be interchanged, and we have

$$n(u,B) - v(u) = \frac{1}{2\pi}\frac{\partial}{\partial u}\int_{\partial D_u} \log \frac{|\Lambda(\zeta), B|}{|\Lambda(\zeta)|\cdot|B|} \; d^c\tau,$$

where $\zeta \in \partial D_u$.

We put

$$N(u,B) = \int_0^u n(t,B)dt, \tag{9.18}$$

(9.19) $$T(u) = \int_0^u v(t)dt .$$

The function $T(u)$ is called the *order function* or the *characteristic function* of Nevanlinna. Integrating the above equation with respect to u, we get the theorem (the integration needs justification; cf. [13]):

(B) (Integrated form of the first main theorem). Let $f: M \to Gr(N,k)$ be a holomorphic curve, where M is a Riemann surface with an exhaustion function τ. Let $\gamma(u)$ be the real curve defined by $\tau = u$. Let B be a fixed $(k+1)$-vector such that $f(\gamma(u)) \cap \Sigma_B = \emptyset$. Introduce the function

$$m(u,B) = \frac{-1}{2\pi} \int_{\gamma(u)} \log \frac{|\Lambda(\zeta),B|}{|\Lambda(\gamma)||B|} d^c\tau , \qquad \zeta \in \gamma(u) .$$

Then

(9.21) $$N(u,B) - T(u) = - m(u,B) + m(0,B) .$$

The function $m(u,B)$ is called the *compensating function* with respect to B. In fact, (9.21) can be written

(9.21a) $$N(u,B) + m(u,B) = T(u) + m(0,B) .$$

Thus $m(u,B)$ measures the "deficiency" of $N(u,B)$ from the order function $T(u)$.

By the Schwarz inequality (8.32) we have

$$\log \frac{|\Lambda(\zeta),B|}{|\Lambda(\zeta)|\cdot|B|} \leqq 0 .$$

By (9.14) we see that $d^c\tau$ is positive along $\gamma(u)$. It follows that

$$m(u,B) \geqq 0 .$$

From (9.21a) we get the inequality

$$N(u,B) \leqq T(u) + m(0,B) , \tag{9.22}$$

which expresses the remarkable fact that $T(u)$ is an upper bound for $N(u,B)$ for all B, $m(0,B)$ being a constant relative to u.

The classical problem of value distribution of meromorphic functions in the sense of Picard-Borel-Nevanlinna can be generalized to holomorphic curves in a Grassmann manifold. We see from the Corollary to Theorem (A) that if M is compact and without boundary and if f is not a constant map, then $f(M)$ meets every Σ_B. The question is whether a similar statement can be made on the equidistribution of $f(M)$ relative to all the Σ_B when M is non-compact but is "conformally large."

The inequality (9.22) furnishes a key to such results. To make the main ideas clear we restrict ourselves to the classical case of a holomorphic mapping $f: M \to P_1$, where

$$M = C_0 = \{z \in C \,\big|\, |z| > 1\} . \tag{9.23}$$

In other words, C_0 is the complex line with a unit disk removed. As its exhaustion function we take $\tau = \log r > 0$, $z = re^{i\theta}$ We suppose that f is not a constant map.

We will derive from (9.22) the theorem that $f(C_0)$ is dense in P_1. In fact, suppose that $P_1 - \overline{f(C_0)} \neq \emptyset$. Let dB be the element of area in P_1 so normalized that

$$\int_{P_1} dB = 1 . \tag{9.24}$$

Let A be the area of $\overline{f(C_0)}$, so that $A < 1$. We integrate the inequality (9.22) with respect to dB over $\overline{f(C_0)}$. Since $v(u)$ is the area of the image $f(D_u)$ (cf. (9.15)), we have

$$\int_{\overline{f(C_0)}} n(u,B)dB = v(u) ,$$

and the integration gives

$$T(u) < AT(u) + \text{const.}$$

Since $T(u) \to \infty$ as $u \to \infty$, this leads to a contradiction.

A deeper study of the value distribution of the meromorphic function f or, what is the same, of the equidistribution of the image set $f(C_0)$ consists in the investigation of the relation of $f(C_0)$ with respect to a given set $A_i \in P_1$, $1 \leqq i \leqq s$, of mutually distinct points. An idea originating from F. Nevanlinnna and Ahlfors is to integrate the inequality (9.22) over a density with singularities at the points A_i. Let

$$(9.25) \qquad \rho(B) = c \prod_i \left(\frac{|B,A_i|}{|B|\,|A_i|} \right)^{-2\lambda} , \quad 0 < \lambda < 1 ,$$

where the constant c is so chosen that

$$(9.26) \qquad \int_{P_1} \rho(B)dB = 1 .$$

Under the mapping f we have

$$(9.27) \qquad f^*dB = \sigma^2 r\, dr \wedge d\theta ,$$

so that σ^2 is the ratio of the elements of area. σ is zero at the points where the linear mapping induced by f on the tangent spaces is not an isomorphism, i.e., at the branch points of the meromorphic function f. Integration of (9.22) over P_1 gives the inequality

$$\int_0^u du \int_1^{\exp u} r\,dr \int_0^{2\pi} \rho\sigma^2 d\theta < T(u) + \text{const.} \tag{9.28}$$

The inequality (9.28) has an implication given by the lemma:

(C) Suppose (9.28) is valid. Then

$$\frac{1}{2\pi}\int_0^{2\pi} \log(\rho\sigma^2)d\theta \quad < \quad \kappa^2 \log T(u) - 2u + \text{const.}, \qquad \kappa > 1 , \tag{9.29}$$

with the exception of a subset E of $u \in R^+$ such that $\int_E du < \infty$.

To prove this let $F(r)$, $G(r)$ be positive-valued functions such that $F(r)$ is of class C^1 and increasing. Suppose that

$$F'(r) \quad > \quad F(r)^\kappa\, G(r) ,$$

where $\kappa > 1$ is a constant. Integration gives

$$\frac{1}{\kappa-1}\,(F^{-\kappa+1}(1) - F^{-\kappa+1}(r)) \quad > \quad \int_1^r G(r)dr ,$$

so that

$$\int_1^\infty G(r)dr \quad < \quad +\infty \tag{9.30}$$

Thus with the exception of a set of r-intervals for which (9.30) holds we have

$$F'(r) \quad \leqq \quad F^\kappa(r)\, G(r) ,$$

and hence

$$\log F' \quad \leqq \quad \kappa \log F + \log G .$$

Successive applications of this formula give respectively

$$\log \int_0^{2\pi} \rho\sigma^2 \, d\theta + \log r \leqq \kappa \log \left\{ \int_1^r r \, dr \int_0^{2\pi} \rho\sigma^2 \, d\theta \right\} + \log G ,$$

$$\log \left\{ \int_1^r r dr \int_0^{2\pi} \rho\sigma^2 \, d\theta \right\} - \log r \leqq \kappa \log T(u) + O(1) + \log G .$$

Combining them we get

$$\log \int_0^{2\pi} \rho\sigma^2 \, d\theta \leqq \kappa^2 \log T(u) + (\kappa-1) \log r + (\kappa+1) \log G + O(1) .$$

Choosing $G(r) = 1/r$ and using the concavity of the logarithmic function

$$(9.31) \qquad \frac{1}{2\pi} \int_0^{2\pi} \log(\rho\sigma^2) d\theta \leqq \log \left\{ \frac{1}{2\pi} \int_0^{2\pi} \rho\sigma^2 \, d\theta \right\} ,$$

we get (9.29).

To draw meaningful conclusions from the inequality (9.29) we need an estimate for the integral

$$\frac{1}{2\pi} \int_0^{2\pi} \log \sigma \, d\theta .$$

For such an estimate consider the projective line $P_1 = Gr(1,0)$. Let $\Lambda = (1,t)$ be its homogeneous coordinate vector so that $0 \leqq |t| \leqq \infty$, including infinity. By (8.43) its metric is

$$(9.32) \qquad ds^2 = \frac{dt \, d\bar{t}}{(1+t\bar{t})^2}$$

By (8.46) its Kähler form is

$$\hat{H}_0 = \frac{i}{2} \frac{dt \wedge d\bar{t}}{(1+t\bar{t})^2} = \frac{i}{2} \partial\bar{\partial} \log(1+t\bar{t}) ,$$

as can be directly verified. A direct computation gives

$$(9.33) \qquad \int_{P_1} \hat{H}_0 = \pi$$

so that

$$(9.34) \qquad dB = \frac{i}{2\pi} \partial\bar{\partial} \log(1+t\bar{t}) .$$

Under the mapping f, t is a holomorphic function of z and we have by definition

$$\sigma = (1+t\bar{t})^{-1} \pi^{-1/2} \left|\frac{\partial t}{\partial z}\right|$$

It follows that

$$f^*dB = -\frac{1}{4\pi i} d(\partial-\bar{\partial}) \log \sigma , \qquad \sigma \neq 0 .$$

Applying Stokes' theorem to the domain D_u and supposing that ∂D_u contains no branch point, we get

$$(9.35) \qquad -2v(u) = \frac{1}{2\pi i} \int_{\partial D_u} (\partial-\bar{\partial}) \log \sigma - w(u) ,$$

where $w(u)$ is the sum of the orders of the branch points in D_u, necessarily finite in number (D_u being compact). Integrating with respect to u, we get

$$(9.36) \qquad -2T(u) = \frac{1}{2\pi} \int_0^{2\pi} \log \sigma \, d\sigma - W(u) + \text{const.}$$

where

$$(9.37) \qquad W(u) = \int_0^u w(u) du .$$

We have thus the inequality

(9.38) $$\frac{1}{2\pi}\int_0^{2\pi} \log \sigma \, d\theta > -2T(u) + \text{const.}$$

By (9.25), (9.20), and (9.38) we get

(9.39) $$\frac{1}{2\pi}\int_0^{2\pi} \log(\rho\sigma^2)d\theta > \text{const.} + 2\lambda \sum_i m(u,A_i) - 4T(u) .$$

We introduce the defect of the point $A \in P_1$ by

(9.40) $$\delta(A) = \liminf \frac{m(u,A)}{T(u)} \quad \text{as} \quad u \to \infty .$$

Thus $\delta(A) = 1$ if $A \notin f(C_0)$ (cf. (9.21)). By letting $\lambda \to 1$ in (9.39), we immediately get, by using (9.29), the theorem:

(D) (Nevanlinna's defect relation.) Let $f: C_0 \to P_1$ be a non-constant holomorphic mapping. Let A_i , $1 \leqq i \leqq s$, be a set of mutually distinct points of P_1 . Then

(9.41) $$\sum_i \delta(A_i) \leqq 2 .$$

Corollary. (Picard's Theorem) A non-constant meromorphic function in C omits at most two values.

BIBLIOGRAPHY

I. Books

1. R. L. Bishop and R. J. Crittenden, *Geometry of Manifolds*, Academic Press, New York, 1964.

2. R. Godement, *Théorie des faisceaux, Actualités Sci. et Indus.* No. 1252, Hermann, Paris, 1958.

3. S. I. Goldberg, *Curvature and Homology*, Academic Press, New York, 1962.

4. R. C. Gunning, *Lectures on Riemann Surfaces*, Princeton Univ. Press, 1966.

5. R. C. Gunning and H. Rossi, *Analytic functions of several complex variables*, Prentice Hall, Englewood Cliffs, N. J., 1965.

6. N. J. Hicks, *Notes on Differential Geometry*, Math. Studies No. 3, van Nostrand, Princeton, New Jersey, 1965.

7. F. Hirzebruch, *Topological Methods in Algebraic Geometry*, Springer, 1966.

8. W. V. D. Hodge and D. Pedoe, *Methods of Algebraic Geometry*, Vol. 2, Cambridge Univ. Press, 1952.

9. S. T. Hu, *Homology Theory*, Holden Day, San Francisco, 1966.

10. G. de Rham, *Variétés Différentiables*, Hermann, Paris, 1955.

11. N. Steenrod, *The Topology of Fibre Bundles*, Princeton Univ. Press, 1951.

12. A. Weil, *Introduction à l'Etude des Variétés Kähleriennes*, Hermann, Paris, 1958.

13. H. Weyl, *Meromorphic Functions and Analytic Curves*, Annals of Math. Studies, No. 12, Princeton Univ. Press, 1943.

II. Articles

14. M. F. Atiyah, "Some examples of complex manifolds," Bonn. Math. Schr 6 (1958), 1-28.

15. R. Bott and S. S. Chern, "Hermitian vector bundles and the equidistribution of the zeroes of their holomorphic sections, Acta Math. 114 (1965), 71-112.

16. S. S. Chern, "An elementary proof of the existence of isothermal parameters on a surface," Proc. Amer. Math. Soc. 6 (1955), 771-782.

17. S. S. Chern, "The geometry of G-structures," Bull. Amer. Math. Soc. 72 (1966), 167-219.

18. K. Kodaira, "On Kähler varieties of restricted type," Ann. of Math. 60 (1954), 28-48.

19. K. Kodaira and D. C. Spencer, "Groups of complex line bundles over compact Kähler varieties; Divisor class groups on algebraic varieties," Proc. Nat. Acad. Sci. 39 (1953), 868-877.

20. A. Newlander and L. Nirenberg, "Complex analytic coordinates in almost complex manifolds," Ann. of Math. 65 (1957), 391-404.

21. A. van de Ven, "On the Chern numbers of certain complex and almost complex manifolds," Proc. Nat. Acad. Sci. 55 (1966), 1624-1627.

III. Added During Second Edition

22. P. A. Griffiths, *Entire Holomorphic Mappings in One and Several Complex Variables*, Ann. of Math Studies 85, Princeton Univ. Press, 1976.

23. P. A. Griffiths and J. Harris, *Principles of Algebraic Geometry*, Wiley, 1978.

24. A. Lascoux and M. Berger, *Variétés Kähleriennes Compactes*, Springer Lecture Notes No. 154, 1970.

25. J. Milnor, *Morse Theory*, Ann. of Math. Studies 51, Princeton Univ. Press, 1963.

26. J. Milnor and J. Stasheff, *Lectures on Characteristic Classes*, Ann. of Math. Studies 76, Princeton Univ. Press, 1974.

27. J. Morrow and K. Kodaira, *Complex Manifolds*, Holt, Rinehart, and Winston, New York, 1971.

28. W. Stoll, *Invariant Forms on Grassmann Manifolds*, Ann. of Math Studies 89, Princeton Univ. Press, 1977.

29. R. O. Wells, Jr., *Differential Analysis on Complex Manifolds*, Prentice Hall, 1973.

Appendix: Geometry of Characteristic Classes[1]

1. Historical Remarks and Examples

The last few decades have seen the development, in different branches of mathematics, of the notion of a local product structure, i.e., fiber spaces and their generalizations. Characteristic classes are the simplest global invariants which measure the deviation of a local product structure from a product structure. They are intimately related to the notion of curvature in differential geometry. In fact, a real characteristic class is a "total curvature," according to a well-defined relationship. We will give in this paper an exposition of the relations between characteristic classes and curvature and discuss some of their applications.

The simplest characteristic class is the Euler characteristic. If M is a finite cell complex, its Euler characteristic is defined by

$$\chi(M) = \sum_k (-1)^k \alpha_k = \sum_i (-1)^k b_k , \tag{1}$$

where α_k is the number of k-cells and b_k is the k-dimensional Betti number of M. The equality of the last two expressions in (1) is known as the Euler-Poincaré formula.

Now let M be a compact oriented differentiable manifold of dimension n and let ξ be a smooth vector field on M with isolated zeroes. Each zero can be assigned a multiplicity. In his dissertation (1927) H. Hopf proved that

[1] Reprinted by permission from Proc. 13th Biennial Seminar, Canadian Math. Congress, 1972.

(2) $$\chi(M) = \sum \text{ zeroes of } \xi \,.$$

This gives a differential topological meaning to $\chi(M)$.

This idea can be immediately generalized. Instead of one vector field we consider k smooth vector fields $\xi_1,\ldots,\xi_k$. In the generic case the points on M where the exterior product $\xi_1 \wedge \ldots \wedge \xi_k = 0$, i.e., where the vectors are linearly dependent, form a (k-1)-dimensional submanifold. Depending on the parity of $n-k$, this defines a (k-1)-dimensional cycle, with integer coefficients Z or with coefficients Z_2, whose homology class, and in particular the homology class mod 2 in all cases, is independent of the choice of the k vector fields. Because the linear dependence of vector fields is expressed by "conditions," it is more proper to define the differential topological invariants so obtained as cohomology classes. This leads to the Stiefel-Whitney cohomology classes $w^i \in H^i(M,Z_2)$, $1 \leqq i \leqq n-1$, $i = n-k+1$. The nth Stiefel-Whitney class corresponding to $k = 1$ or the Euler class has integer coefficients $w^n \in H^n(M,Z)$. It is related to $\chi(M)$ by

(3) $$\chi(M) = \int_M w^n ,$$

where we write the pairing of homology and cohomology by an integral.

Whitney went much farther. He saw the great generality of the notion of a vector bundle over an arbitrary topological space M. (Actually Whitney considered sphere bundles, thus gaining the advantage that the fibers are compact but losing the linear structure on the fibers. He was not concerned with the latter, as he was only interested in topological problems.) He also saw the effectiveness of the principal bundles and the fact that the universal principal bundle

(4) $$O(q+N)/O(N) \longrightarrow O(q+N)/\{O(q) \times O(N)\} = G(q,N) ,$$

say, has the property

(5) $$\pi_i(O(q+N)/O(N)) = 0 , \quad 0 \leqq i < N ,$$

where π_i is the ith homotopy group. The left-hand side of (4) is called a Stiefel manifold and can be regarded as the space of all orthonormal q-frames through a fixed point 0 of the euclidean space E^{q+N} of dimension q+N and the right-hand side is the Grassmann manifold of all q-dimensional linear spaces through 0 in E^{q+N}, while the mapping π in (4) can be interpreted geometrically as taking the q-dimensional space spanned by the q vectors of the frame. Thus the universal principal bundle has the feature that its total space has a string of vanishing homotopy groups while its base space, the Grassmann manifold, has rich homological properties. The associated sphere bundle of the principal bundle (4) can be written

(6) $$O(q+N)/\{O(q-1) \times O(N)\} \to O(q+N)/\{O(q) \times O(N)\} .$$

The importance of the universal bundle lies in the Whitney-Pontrjagin imbedding theorem: let M be a finite cell complex. A sphere bundle of fiber dimension q-1 (or a vector bundle E of fiber dimension q) over M can be induced by a continuous mapping $f: M \to G(q,N)$, $\dim M < N$, and f is defined up to a homotopy.

Let $u \in H^i(G(q,N),A)$ be a cohomology class with coefficient group A. It follows from the above theorem that $f^*u \in H^i(M,A)$ depends only on the bundle. It is called a *characteristic class corresponding to the universal class u.*

Example 1. Consider all the q-dimensional linear spaces X through 0 in E^{q+N} satisfying the Schubert condition

(7) $$\dim(X \cap E^{i+N-1}) \geqq i , \quad 1 \leqq i \leqq q ,$$

where E^{i+N-1} is a fixed space of dimension $i + N - 1$ through 0. They form a cycle mod 2 of dimension $qN - i$ in $G(q,N)$. The dual of its homology class is an element $\tilde{w}^i \in H^i(G(q,N),Z_2)$ and is called the ith universal Stiefel-Whitney class. Its image $w^i(E) = f^*\tilde{w}^i \in H^i(M,Z_2)$, $1 \leqq i \leqq q$, is called the Stiefel-Whitney class of the bundle E.

Example 2. Similarly, consider the q-dimensional linear spaces X through 0 satisfying the condition

$$\dim(X \cap E^{2k+N-2}) \geqq 2k , \tag{8}$$

where E^{2k+N-2} is fixed, with its superscript indicating the dimension. They form a cycle of dimension $qN-4k$ with integer coefficients. The dual of its homology class is an element $\tilde{p}_k \in H^{4k}(G(q,N),Z)$ and is called a universal Pontrjagin class. Its image $p_k(E) = f^*p_k \in H^{4k}(M,Z)$, $1 \leqq k \leqq \left[\frac{n}{4}\right]$, $n = \dim M$, is called a Pontrjagin class of E.

Example 3. It has been known that the complex Grassmann manifold

$$G(q,N,C) = U(q+N)/U(q) \times U(N) \tag{9}$$

has simpler topological properties than the real ones. In fact, it is simply connected, has no torsion (i.e., no homology class of finite order), and its odd-dimensional homology classes are all zero. $G(q,N,C)$ can be regarded as the manifold of all q-dimensional linear spaces X through a fixed point 0 in the complex number space C_{q+N} of dimension $q + N$. Imitating Example 1, let C_{i+N-1} be a fixed space of dimension $i + N - 1$ through 0. Then all the X satisfying the condition

$$\dim(X \cap C_{i+N-1}) \geqq i , \qquad 1 \leqq i \leqq q , \tag{10}$$

form a cycle of real dimension $2(qN-i)$ with coefficients Z. As above, this defines the Chern classes $c_i(E) \in H^{2i}(M,Z)$, $1 \leqq i \leqq q$, of a complex vector bundle E and they are cohomology classes with integer coefficients.

When applied to the tangent bundle of a differentiable manifold the Stiefel-Whitney classes and the Pontrjagin classes are invariants of the differentiable structure. Similarly, the Chern classes of the tangent bundle of a complex manifold are invariants of the complex structure.

It is of great importance to know whether and how the characteristic classes are related to the underlying topological structure of the manifold. The first such relation is the identification of the Euler class with the Euler characteristic, as given by (3). It was proved by Thom and Wu that the Stiefel-Whitney classes can be defined through the Steenrod squaring operations and are topological invariants. On a compact complex manifold of dimension m we have, in analogy to (3).

$$\chi(M) = \int_M c_m(M) , \tag{11}$$

where $c_m(M)$ denotes the mth Chern class of the tangent bundle of M .

From the Pontrjagin classes of the tangent bundle of a compact oriented differentiable manifold M^{4k} of dimension $4k$ Hirzebruch constructed a number called the L-genus and, using Thom's cobordism theory, proved that it is equal to the signature of M^{4k} . In the simplest case $k = 1$ the relation is

$$\operatorname{sign}(M) = \frac{1}{3}\int_M p_1(M) , \quad M = M^4 . \tag{12}$$

In particular, it shows that the integral at the right-hand side is divisible by 3.

The characteristic classes are closely related to the notion of curvature in differential geometry. In this respect one could take as a starting-point the theorem in plane goemetry that the sum of angles of a triangle is equal to π . More generally, let D be a domain in a two-dimensional riemannian manifold, whose boundary ∂D is sectionally smooth. Then its Euler characteristic is given by the Gauss-Bonnet formula

$$(13) \qquad 2\pi\chi(D) = \sum_i (\pi-\alpha_i) + \int_{\partial D} \frac{ds}{\rho_g} + \iint_D K\, dA ,$$

where the first term at the right-hand side is the sum of the exterior angles at the corners, the second term is the integral of the geodesic curvature, and the last term is the integral of the gaussian curvature. They are respectively the point curvature, the line curvature, and the surface curvature of the domain D , and the Gauss-Bonnet formula should be interpreted as expressing the Euler characteristic $\chi(D)$ as a total curvature.

The interpretation has a far-reaching generalization. Let $\pi: E \to M$ be a real (C^∞-differentiable) vector bundle of fiber dimension q . Let $\Gamma(E)$ be the space of sections of E , i.e., smooth mappings $s: M \to E$ such that $\pi \circ s$ = identity. A *connection* or a *covariant differential* in E is a structure which allows the differentiation of sections. It is a mapping

$$(14) \qquad D: \Gamma(E) \to \Gamma(T^* \otimes E) ,$$

where T^* is the cotangent bundle of M and the right-hand side stands for the space of sections of the tensor product bundle $T^* \otimes E$, such that the following two conditions are satisfied:

$$(15a) \qquad D(s_1 + s_2) = Ds_1 + Ds_2 , \qquad s_1, s_2 \in \Gamma(E) ,$$

$$(15b) \qquad D(fs) = df \otimes s + fDs , \qquad s \in \Gamma(E) ,$$

where f in (15b) is a C^∞-function.

Let s_i , $1 \leqq i \leqq q$, be a local frame field, i.e., be q sections defined in a neighborhood, which are everywhere linearly independent. Then we can write

$$Ds_i = \sum_i \theta_i^j \otimes s_j , \tag{16}$$

where $\theta = (\theta_i^j)$, $1 \leqq i,j \leqq q$, is a matrix of one-forms, the *connection matrix.* Putting

$$^t s = (s_1,\ldots,s_q) , \quad ^t s = \text{transpose of } s , \tag{17}$$

we can write (16) as a matrix equation

$$Ds = \theta \otimes s . \tag{16a}$$

The effect on the connection matrix under a change of the frame field can easily be found. In fact, let

$$s' = gs \tag{18}$$

be a new frame field, where g is a nonsingular $(q\times q)$-matrix of C^∞-functions. Let θ' be the connection matrix relative to the frame field s' so that

$$Ds' = \theta' \otimes s' . \tag{19}$$

Using the properties of D as expressed by (15a) and (15b), we find immediately

$$\theta' g = dg + g\theta . \tag{20}$$

This is the equation for the change of the connection matrix under a change of the frame field.

Taking the exterior derivative of (20), we get

$$\Theta' = g\Theta g^{-1} \tag{21}$$

where

(22) $$\Theta = d\theta - \theta \wedge \theta$$

and Θ' is defined in terms of θ' by a similar equation. Θ is a $(q\times q)$-matrix of two-forms and is called the *curvature matrix* relative to the frame field s. Equation (21) shows that it undergoes a very simple transformation law under a change of the frame field. As a consequence it follows from (21) that $tr(\Theta^k)$ is a form of degree $2k$ globally defined in M. Moreover, $tr(\Theta^k)$ can be proved to be a closed form and the cohomology class $\{tr(\Theta^k)\} \in H^{2k}(M,R)$ it represents in the sense of de Rham's theorem can be identified with a characteristic class of E.

Example 1. Let M^4 be a compact oriented differentiable manifold of dimension 4. Let $\Theta = (\Theta_i^j)$, $1 \leq i,j \leq 4$, be the curvature matrix of a connection in the tangent bundle of M^4. Then $p_1(M^4)$ can be identified with a numerical multiple of $\{tr(\Theta^2)\}$. By (12) we will have the integral formula

(23) $$\text{sign}(M) = \frac{1}{24\pi^2}\int_M \sum_{i,j} \Theta_i^j \wedge \Theta_j^i, \qquad M^4 = M.$$

Example 2. When the bundle $\pi: E \to M$ is oriented and has a riemannian structure, the structure group is reduced to $SO(q)$, and we can restrict our consideration to frame fields consisting of orthonormal frames. Then both connection and curvature matrices are anti-symmetric, and we have

(24) $$\Theta = -{}^t\Theta = (\Theta_{ij}), \qquad \Theta_{ij} + \Theta_{ji} = 0.$$

If q is even, the pfaffian

(25)
$$Pf(\Theta) = \frac{(-1)^r}{2^q\pi^r r!} \sum_i \varepsilon_{i_1\ldots i_q}\Theta_{i_1 i_2}\wedge\ldots\wedge\Theta_{i_{q-1}i_q}, \qquad r = q/2,$$

represents the Euler class, i.e.,

$$\{Pf(\Theta)\} = w^q(E) . \tag{26}$$

Formula (26) is essentially the high-dimensional Gauss-Bonnet Theorem.

The starting point of this paper is the Weil homomorphism which gives a representation of characteristic classes with real coefficients by the curvature forms of a connection in the bundle. The connection makes many cochain constructions canonical and gives geometrical meaning to them. The resulting homomorphism exhibits a relationship between local and global properties which is not available in the topological theory of characteristic classes. It is effective when the manifold has more structure, such as a foliated structure (Bott's theorem) or a complex structure with a holomorphic bundle over it. In the latter case we will show the fundamental rôle played by the curvature forms representing characteristic classes in the Ahlfors-Weyl theory of holomorphic curves in complex projective space, which generalizes the theory of value distributions in complex function theory. This is the case of the geometry of a noncompact manifold where deep studies have been carried out.

In another direction the Weil homomorphism leads to new global invariants when certain curvature forms vanish. In recent works of Chern and Simons such invariants are found to be nontrivial global invariants of the underlying conformal or projective structure of a riemannian manifold.

This exposition will be devoted to the following topics:

1. Weil homomorphism;
2. Bott's theorem on foliated manifolds;
3. Secondary invariants (Chern-Simons);
4. Vector fields and characteristic numbers (Bott-Baum-Cheeger);
5. Holomorphic curves (Ahlfors-Weyl).

2. Connections

We will develop the fundamental notions of a connection in a principal bundle with a Lie group as structure group. We begin by a review and an explanation of our notations on Lie groups. All manifolds and mappings are C^∞ .

Let G be a Lie group of dimension r . A left translation $L_a\colon G \to G$ is defined by $L_a\colon s \to as$, $a, s \in G$, a fixed. Let e be the unit element of G and T_e the tangent space at e. A tangent vector $X_e \in T_e$ generates a left-invariant vector field given by $X_s = (L_s)_* X_e$. If T_e^* is the cotangent space at e and $\omega_e \in T_e^*$, we get a left-invariant one-form or Maurer-Cartan form ω_s by the definition

$$(27) \qquad \omega_s = (L_s^{-1})^* \omega_e \qquad \text{or} \qquad L_s^* \omega_s = \omega_e .$$

Let ω_e^i , $1 \leqq i \leqq r$, be a basis in T_e^* . Then $\omega^i = \omega_s^i \in T_s^*$ are everywhere linearly independent and we have

(28)

$$d\omega^i = \frac{1}{2} \sum_{j,k} c^i_{jk} \omega^j \wedge \omega^k , \qquad c^i_{jk} + c^i_{kj} = 0 , \qquad 1 \leqq i,j,k \leqq r .$$

It is easily proved that c^i_{jk} are constants, the constants of structure of G . Equations (28) are known as the Maurer-Cartan structure equations.

Let $X_i = (X_i)_s \in T_s$ be a dual basis to ω^i . The X_i are left-invariant vector fields or, what is the same, linear differential operators of the first order. Dual to (28) are the equations of Lie:

$$(29) \qquad [X_j, X_i] = - \sum_k c^k_{ji} X_k .$$

The tangent space T_e has an algebra structure given by the bracket. It is called the Lie algebra of G and will be denoted by $\mathfrak{g}$.

For a fixed $a \in G$ the inner autmorphism $s \to asa^{-1}$ leaves e fixed and induces a linear mapping

(30) $$\mathrm{ad}(a)\colon \mathfrak{g} \to \mathfrak{g} ,$$

called the adjoint mapping. We have

(31) $$\mathrm{ad}(ab) = \mathrm{ad}(a)\mathrm{ad}(b) , \qquad a,b \in G$$

(32) $$\mathrm{ad}(a)[X,Y] = [\mathrm{ad}(a)X, \mathrm{ad}(a)Y] , \qquad X,Y \in \mathfrak{g} .$$

The first relation is immediate and the second is easy to prove.

Let M be a manifold. It will be desirable to consider $\mathfrak{g}$-valued exterior differential forms in M. As $\mathfrak{g}$ has an algebra structure, such forms can be multiplied. In fact, every $\mathfrak{g}$-valued form is a sum of terms $X \otimes \omega$, where ω is an exterior differential form and $X \in \mathfrak{g}$. We define

(33) $$[X \otimes \omega, Y \otimes \theta] = [X,Y] \otimes (\omega \wedge \theta) .$$

Distributivity in both factors then defines the multiplication of any two $\mathfrak{g}$-valued forms. Interchange of order of multiplication follows the rule

(34) $$[X \otimes \omega, Y \otimes \theta] = (-1)^{rs+1} [Y \otimes \theta, X \otimes \omega] ,$$
$$r = \deg \omega , \qquad s = \deg \theta .$$

This notion allows us to write the Maurer-Cartan equations (28) in a simple form. The expression

(35) $$\omega = \sum_i (X_i)_e \otimes \omega_s^i$$

defines a left-invariant $\mathfrak{g}$-valued one-form in G, which is independent of the choice of the basis. It is *the* Maurer-Cartan form of G. Using (28) and (29) we have

(36) $$d\omega = -\frac{1}{2} [\omega,\omega] .$$

This writes the Maurer-Cartan equation in a basis-free form.

Exterior differentiation of (36) gives the Jacobi identity:

$$[\omega,[\omega,\omega]] = 0 . \tag{37}$$

What we have discussed on left translations naturally holds also for right translations. In particular, we have a right-invariant one-form α in G. Under the mappings $s \to s^{-1}$, $s \in G$, ω goes into $-\alpha$. We derive therefore from (36)

$$d\alpha = \frac{1}{2}[\alpha,\alpha] . \tag{38}$$

If we denote by ds the identity endomorphism in T_s and consider it as an element of $T_s \otimes T_s^*$, then we can write

$$\omega = (L_{s^{-1}})_* ds = s^{-1} ds , \tag{39}$$

where $(L_{s^{-1}})_*$ acts only on the first factor T_s in the tensor product $T_s \otimes T_s^*$; the last expression is a convenient abbreviation. In the same way we can write $\alpha = dss^{-1}$.

<u>Example</u>. $G = GL(q;R)$. We can regard it as the group of all nonsingular $(q\times q)$-matrices X with real elements. Then $\mathfrak{g}$ is the space of all $(q\times q)$-matrices, and $\omega = X^{-1}dX$. Thus the notation in (39) has in this case a concrete meaning. The Maurer-Cartan equation is

$$d\omega = -\omega \wedge \omega . \tag{40}$$

A principal fiber bundle with a group G is a mapping

$$\pi: P \to M , \tag{41}$$

which satisfies the following conditions:

1. G acts freely on P to the left, i.e., there is an action $G \times P \to P$ given by $(a,z) \to az = L_a z \in P$, $a \in G$, $z \in P$, such that $az \neq z$ when $a \neq e$;

2. $M = P/G$;

3. P is locally trivial, i.e., there is an open covering $\{U,V,\ldots\}$ of M such that to each member U of the covering there is a chart $\psi_U\colon \pi^{-1}(U) \to U \times G$, with $\psi_U(z) = (\pi(z) = x, s_U(z))$, $z \in \pi^{-1}(U)$, satisfying

$$s_U(az) = as_U(z) , \qquad z \in \pi^{-1}(U) , \qquad a \in G . \tag{42}$$

Suppose $z \in \pi^{-1}(U \cap V)$. By (42) we have also

$$s_V(az) = as_V(z) ,$$

so that

$$s_U(az)^{-1}s_V(az) = s_U(z)^{-1}s_V(z)$$

is independent of a and depends only on $x = \pi(z)$. We put

$$s_U(z)^{-1}s_V(z) = g_{UV}(x)$$

or

$$s_U g_{UV} = s_V . \tag{43}$$

The g_{UV} are mappings of $U \cap V$ into G and satisfy the relations

$$g_{UV}g_{VU} = e \quad \text{in } U \cap V ,$$
$$g_{UV}g_{VW}g_{WU} = e \quad \text{in } U \cap V \cap W .$$

They are called the transition functions of the bundle. It is well-known that the bundle, the principal bundle or any of its associated bundles, can be constructed from the transition functions.

The bundle structure in P defines in each tangent space T_z a subspace $G_z = \pi_*^{-1}(0)$, called the vertical space. By (43) each fiber of P is the group manifold G defined up to right translations. It is thus meaningful to talk about $\mathfrak{g}$-valued forms in P which restrict to the right-invariant form $ds_U s_U^{-1}$ on a fiber.

We will give three definitions of a connection, which are all equivalent:

First definition of a connection. A *connection* is a C^∞-family of subspaces H_z (the horizontal spaces) in T_z satisfying the conditions:

1. $T_z = G_z + H_z$, $\qquad G_z \cap H_z = 0$;

2. $H_{az} = (L_a)_* H_z$.

The second condition means that the family of horizontal spaces is invariant under the action of the group G .

Second definition of a connection. This is the dual of the first definition, by giving instead of $H_z \in T_z$ its annihilator V_z^* in the cotangent space T_z^* . This in turn is equivalent to giving a $\mathfrak{g}$-valued one-form ϕ in P which restricts to $ds_U s_U^{-1}$ on a fiber, i.e., locally

$$\phi(z) = ds_U s_U^{-1} + \theta_U(x, s_U, dx)$$

such that

$$\phi(az) = \mathrm{ad}(a)\phi(z) \, . \tag{45}$$

The last condition is equivalent to condition (2) in the first definition. It implies that locally

$$\phi(z) = ds_U s_U^{-1} + \mathrm{ad}(s_U)\theta_U(x, dx) \, , \tag{46}$$

where $\theta_U(x, dx)$ is a $\mathfrak{g}$-valued one-form in U . Thus the second definition of a connection is the existence of a $\mathfrak{g}$-valued one-form in P , which has the local expression (46).

Third definition of a connection. When we express the condition that in $\pi^{-1}(U \cap V)$ the right-hand side of (46) is equal to the corresponding expression with the subscript V , we get

$$\theta_U = dg_{UV} g_{UV}^{-1} + \mathrm{ad}(g_{UV})\theta_V \quad \text{in } U \cap V \, , \tag{47}$$

where the first term at the right-hand side is the pull-back of the right-invariant form in G under g_{UV} . Hence a connection in P is given by a $\mathfrak{g}$-valued one-form θ_U in every member U of an open covering $\{U, V, \ldots\}$ of M , such that in $U \cap V$ the equation (47) holds. This is essentially the classical definition of a connection.

We wish to take the exterior derivative of (46). For this purpose we need the following lemma, which is easily proved (and the proof is omitted here):

Lemma. *Let* θ *be a* g*-valued one-form in* U . *Let* $s \in G$ *and let* $\alpha = dss^{-1}$ *be the right-invariant* g*-valued one-form in* G . *Then, in* $U \times G$, *we have*

$$(48) \qquad d(ad(s)\theta) = ad(s)d\theta + [ad(s)\theta,\alpha] .$$

We put

$$(49) \qquad \Theta_U = d\theta_U - \frac{1}{2}[\theta_U,\theta_U] ,$$

$$(50) \qquad \Phi = d\phi - \frac{1}{2}[\phi,\phi] .$$

Applying the lemma we get by exterior differentiation of (46),

$$(51) \qquad \Phi = ad(s_U)\Theta_U .$$

Thus Φ is a g-valued two-form in P ,which has the local expression (51). Alternately, we have, in $U \cap V$,

$$(52) \qquad \Theta_U = ad(g_{UV})\Theta_V .$$

Either Φ or Θ_U will be called the *curvature form* of the connection.

Exterior differentiation of (50) gives the *Bianchi Identity:*

$$(53) \qquad d\Phi = -[\Phi,\phi] = [\phi,\Phi] .$$

One of the most important cases of this general theory is when $G = GL(q;R)$. As discussed above, s_U is now a nonsingular $(q\times q)$-matrix, θ_U , ϕ are matrices of one-forms, and Θ_U , Φ are matrices of two forms. Equation (46) becomes a matrix equation

$$(54) \qquad \phi = (ds_U + s_U\theta_U)s_U^{-1} .$$

Let σ_U (resp. σ_V) be the one-rowed matrix formed by the first row of s_U (resp. s_V). Then (43) gives, by taking the first rows of both sides,

$$\sigma_U g_{UV} = \sigma_V . \tag{55}$$

This is the equation for the change of chart of the associated vector bundle E, defined as the bundle of the first row vectors of the matrices representing the elements of $GL(q;R)$. Moreover, equating the right-hand side of (54) with the corresponding expression with the subscript V, we get

$$(ds_U + s_U\theta_U)g_{UV} = ds_V + s_V\theta_V . \tag{56}$$

On taking the first rows of both sides of (56), we have

$$D\sigma_U g_{UV} = D\sigma_V , \tag{57}$$

where we put

$$D\sigma_U = d\sigma_U + \sigma_U\theta_U . \tag{58}$$

Applying to a section of E, we can identify this with the operator D in (14). Thus we have shown that the connection in a vector bundle defined in §1 is included as a special case of our general theory.

Another important case is the bundle (4) discussed in §1, which is a principal bundle with the group $O(q)$. This bundle plays a fundamental rôle in the study of submanifolds in euclidean space. As remarked above, its importance in bundle theory arises from the fact that it is a universal bundle when N is large. We will describe a canonical connection in it. Let E^{q+N} be the euclidean space of dimension $q + N$. Let

$$e_A = (e_{A1},\dots,e_{A,q+N}) , \qquad 1 \leqq A,B,C \leqq q + N ,$$

be an orthonormal frame, so that the matrix

$$X = (e_{AB}) \tag{59}$$

is orthogonal. $O(q+N)$ can be identified with the space of all orthonormal frames e_A (or all orthogonal matrices X). Let

$$de_A = \sum_B \alpha_{AB} e_B . \tag{60}$$

Then, if $\alpha = (\alpha_{AB})$, we have

$$\alpha = dXX^{-1} = -{}^t\alpha . \tag{61}$$

The Stiefel manifold $O(q+N)/O(N)$ can be identified with the manifold of all orthonormal frames $e_1, \ldots e_q$ and the Grassmann manifold $O(q+N)/\{O(q) \times O(N)\}$ with the q-planes spanned by $e_1, \ldots, e_q$. The matrix

$$\alpha = (\alpha_{ij}) , \quad 1 \leq i,j \leq q , \tag{62}$$

defines a connection in the bundle (4), as easily verified.

3. Weil Homomorphism

The local expression (51) of the curvature form Φ prompts us to introduce functions $F(X_1, \ldots, X_h)$, $X_i \in g$, $1 \leq i \leq h$, which are real or complex valued and satisfy the conditions:

1. F is h-linear and remains unchanged under any permutation of its arguments;

2. F is "invariant," i.e.,

$$F(ad(a)X_1, \ldots, ad(a)X_h) = F(X_1, \ldots, X_h) , \qquad \text{all } a \in G . \tag{63}$$

To the h-linear function $F(X_1, \ldots, X_h)$ there corresponds the polynomial

$$F(X) = F(X, \ldots, X) , \quad X \in g , \tag{64}$$

of which $F(X_1, \ldots, X_h)$ is the complete polarization. We will call $F(X)$ an invariant polynomial. All invariant polynomials under G form a ring, to be denoted by $I(G)$.

The invariance condition (63) implies its "infinitesimal form"

(65) $$\sum_{1\leq i\leq h} F(X_1,\ldots,[Y,X_i],\ldots,X_h) = 0 \qquad Y,X_i \in g .$$

More generally, if Y is a g-valued one-form and X_i is a g-valued form of degree m_i, $1 \leq i \leq h$, we have

(66) $$\sum_{1\leq i\leq h} (-1)^{m_1+\ldots+m_{i-1}} F(X_1,\ldots,[Y,X_i],\ldots,X_h) = 0 .$$

It follows from (51) that if F is an invariant polynomial of degree h , we have the form of degree 2h :

(67) $$F(\Phi) = F(\Theta_U) .$$

The left-hand side shows that it is globally defined in P, while the right-hand side shows that it is a form in M. Moreover, by the Bianchi identity (53) and by (66), we have

$$\frac{1}{h} df(\Phi) = F([\phi,\Phi],\Phi,\ldots,\Phi) = 0 .$$

Hence $F(\Phi)$ is closed and its cohomology class $\{F(\Phi)\}$ is an element of $H^{2h}(M,R)$. We shall prove that this class depends only on F and is independent of the choice of the connection.

<u>Lemma 3.1</u>. *Let* ϕ_0, ϕ_1 *be* g*-valued one-forms and let* $F \in I(G)$ *be an invariant polynomial of degree* h. *Let*

(68) $$\phi_t = \phi_0 + t\alpha , \qquad \alpha = \phi_1 - \phi_0 ,$$

(69) $$\Phi_t = d\phi_t - \frac{1}{2}[\phi_t,\phi_t] .$$

Then

(70) $$F(\Phi_1) - F(\Phi_0) = hd\int_0^1 F(\alpha,\Phi_t,\ldots,\Phi_t)dt .$$

To prove the lemma we first find

$$\Phi_t = \Phi_0 + t(d\alpha - [\phi_0,\alpha]) - \frac{1}{2} t^2[\alpha,\alpha] .$$

Therefore we have

$$\frac{1}{h}\frac{d}{dt} F(\Phi_t) = F(d\alpha - [\phi_t,\alpha], \Phi_t,\ldots,\Phi_t) .$$

On the other hand,

$$dF(\alpha,\Phi_t,\ldots,\Phi_t) = F(d\alpha,\Phi_t,\ldots,\Phi_t) - (h-1)F(\alpha,[\phi_t,\Phi_t],\Phi_t,\ldots,\Phi_t) .$$

The invariance of F implies, by (66),

$$F([\phi_t,\alpha],\Phi_t,\ldots,\Phi_t) - (h-1)F(\alpha,[\phi_t,\Phi_t],\Phi_t,\ldots,\Phi_t) = 0 .$$

It follows that

$$\frac{1}{h}\frac{d}{dt} F(\Phi_t) = dF(\alpha,\Phi_t,\ldots,\Phi_t) , \tag{71}$$

and the lemma follows by integrating this equation with respect to t.

Corollary 3.1. *Let* ϕ_0, ϕ_1 *be two connections in the bundle* $\pi\colon P \to M$ *and let* $F \in I(G)$. *Then* $F(\Phi_0)$ *and* $F(\Phi_1)$, *as closed forms in* M , *are cohomologous in* M .

Corollary 3.2. *Let* ϕ *be a connection in the bundle* $\pi\colon P \to M$ *and let* $F \in I(G)$. *Then* $F(\Phi)$ *is a coboundary in* P . *More precisely, let*

$$\Phi_t = t\, d\phi - \frac{1}{2} t^2[\phi,\phi] = t\Phi + \frac{1}{2}(t-t^2)[\phi,\phi] . \tag{72}$$

Then

$$F(\Phi) = hd\int_0^1 F(\phi,\Phi_t,\ldots,\Phi_t)dt .$$

By putting

$$w(F) = \{F(\Phi)\} , \qquad F \in I(G) , \tag{74}$$

where the right-hand side denotes the cohomology class represented by the closed form $F(\Phi)$, we have defined a mapping

$$w\colon I(G) \to H^*(M,R) . \tag{75}$$

It is clearly a ring homomorphism and is called the *Weil homomorphism.*

In the case that G is a compact connected Lie group, the Weil homomorphism has a simple geometric interpretation, which we will state without proof (cf. [7], [23]). There is a universal principal bundle $\pi_0 \colon E_G \to B_G$ with group G such that we have the bundle map

(76)
$$\begin{array}{ccc} P & \xrightarrow{f_!} & E_G \\ {\scriptstyle\pi}\downarrow & & \downarrow{\scriptstyle\pi_0} \\ M & \xrightarrow{f} & B_G \end{array}$$

where f is defined up to a homotopy. B_G is called the classifying space with the group G. The following diagram is commutative:

(77)
$$\begin{array}{ccc} I(G) & \xrightarrow{w} & H^*(M,R) \\ & {\scriptstyle w_0}\searrow \quad \nearrow{\scriptstyle f^*} & \\ & H^*(B_G,R) & \end{array} ,$$

and w_0 is an isomorphism. In other words, the invariant polynomials can be identified with the cohomology classes of the classifying space and the Weil homomorphism gives the representatives of characteristic classes by closed differential forms constructed from the curvature forms of a connection.

We put

(78)
$$TF(\phi) = h \int_0^1 F(\phi, \Phi_t, \ldots, \Phi_t)\,dt ,$$

so that (73) can be written

(79)
$$\pi^* F(\Theta_U) = F(\Phi) = d(TF(\phi)) .$$

T will be called the *transgression operator;* it enables $F(\Phi)$ to be written as a coboundary in a canonical way, by the use of a connection.

One application of the transgression operator is the following description of the de Rham ring of P (theorem of Chevalley):

Let G *be a compact connected semi-simple group of rank* r *(= dimension of maximal torus in* G*). Let* $\pi \colon P \to M$ *be a principal*

G-bundle over a compact manifold M *and* ϕ *a connection in* P . *Then the ring* $I(G)$ *of invariant polynomials is generated by elements* $F_1,\dots,F_r$ *and the de Rham ring of* P *can be given as the quotient ring*

$$H^*(P,R) = A/dA , \tag{80}$$

where

$$A = O(TF_1(\phi),\dots,TF_r(\phi)) \tag{81}$$

is the ring of polynomials in $TF_1(\phi),\dots,TF_r(\phi)$ *with coefficients which are forms in* M .

For geometrical applications we will describe in detail the Weil homomorphsim for some of the classical groups [24]:

1. $G = GL(q;C) = \{X|\det X \neq 0\}$, where X is a $(q\times q)$-matrix with complex elements. The coefficients $F_i(X)$, $1 \leqq i \leqq q$, in the polynomial in t :

$$\det\left(tI_q + \frac{i}{2\pi} X\right) = t^q + F_1(X)t^{q-1} + \dots + F_q(X) , \tag{82}$$

where I_q is the $(q\times q)$-unit matrix, are invariant polynomials.

Suppose $\pi: E \to M$ be a complex vector bundle and ϕ be a connection, with the curvature form Φ , so that Φ is a matrix of two-forms. Then we have

$$F_i(\Phi) = c_i(E) \in H^{2i}(M,Z) . \tag{83}$$

Notice that the coefficients are here so chosen that the corresponding classes have integer coefficients.

By the above Corollary 3.1 it suffices to establish this result in the classifying space $B_G = G(q,N;C)$ (N sufficiently large), with its connection defined in a similar way as the one in §2 for the real Grassmann manifold. In other words it is sufficient to consider the universal bundle with its universal connection. The same remark applies in the identification in the next two cases.

2. $G = GL(q;R) = \{X | \det X \neq 0\}$, where X is a $(q\times q)$-matrix with real elements. We put

$$\det\left(tI_q - \frac{1}{2\pi} X\right) = t^q + E_1(X)t^{q-1} + \ldots + E_q(X) . \tag{84}$$

Let $\pi: E \to M$ be a real vector bundle and ϕ be a connection, with the curvature form Φ . Then $\{E_{2k+1}(\Phi)\} = 0$ and

$$\{E_{2k}(\Phi)\} = p_k(E) \in H^{4k}(M,Z) \tag{85}$$

is the kth Pontrjagin class of E.

3. $G = SO(2r)$. A representative of the Euler class was given by formula (26), §1.

As an application of the representation of characteristic classes by curvature forms we will prove a theorem of Bott on foliations [10].

Let M be a manifold of dimension n and TM its tangent bundle. Suppose that TM has a k-dimensional subbundle W , i.e., a smooth family of k-dimensional subspaces $W_x \subset T_x$, $x \in M$, where T_x is the tangent space of M at x . To W_x corresponds the annihilator $W_x^{\perp}$ of dimension $n - k$ in the cotangent space T_x^* at x . The subbundle W is called *integrable*, if there exist local coordinates x^{α} , x^{λ} , $1 \leq \alpha \leq k$, $k + 1 \leq \lambda \leq n$, such that $W_x^{\perp}$ is spanned by dx^{λ} . In other words, the W_x are tangent to the submanifolds of dimension k defined by $x^{\lambda} = \text{const}$. An integrable subbundle is called a *foliation.* The local coordinates with the above properties are defined up to a transformation

$$x^{\alpha} = x^{\alpha}(x'^{\beta},x'^{\mu}) , \quad x^{\lambda} = x^{\lambda}(x'^{\mu}) , \quad 1 \leq \alpha,\beta \leq k, \ k+1 \leq \lambda,\mu \leq n , \tag{86}$$

where the last $n - k$ coordinates transform among themselves. As a consequence the quotient bundle TM/W has as transition functions the

Jacobian matrices $(\partial x^{\lambda}/\partial x'^{\mu})$.

From the existence of the foliation on M we have an ideal F in the ring of exterior differential polynomials, which is generated by dx^{λ} . F is stable under d , i.e., $\alpha \in F$ implies $d\alpha \in F$. Moreover, the product of any $n - k + 1$ elements of F is zero.

From (47) we see by a stepwise extension argument that TM/W has a connection whose connection forms belong to F. By (49) its curvature forms also belong to F. It follows that if F is an invariant polynomial of degree $h > n - k$, the form $F(\Phi) = 0$. Hence every Pontrjagin class, being a polynomial of the Pontrjagin classes defined in §1, of the bundle TM/W is zero, if its dimension is $> 2(n-k)$. We state this theorem of Bott as follows:

Let M *be a compact manifold of dimension* n *, which has a foliation* W *of dimension* k *. Then every Pontrjagin class (with real coefficients) of dimension* $>2(n-k)$ *of the quotient bundle* TM/W *is zero.*

The theorem is remarkable because the integrability of W involves differential conditions, so that it cannot be proved by standard methods in fiber bundles. The necessary condition asserted in the theorem is not vacuous. For example, there are real codimension two subbundles in the complex projective space $P_5(C)$ of complex dimension 5, which do not satisfy the above condition.

4. Secondary Invariants[1]

When the characteristic classes are given representatives by differential forms, the vanishing of the forms leads to further invariants which deserve investigation. We follow the notations of the last section and consider the formula (79). If $F(\Phi) = 0$,

[1]The results in this section are taken from joint work with James Simons, cf. [18].

the form $TF(\phi)$ is closed and defines an element of $H^{2h-1}(P,R)$. The latter depends on the connection ϕ and it is desirable to study the effect on $TF(\phi)$ under a change of the connection. This is given by the following lemma:

<u>Lemma 4.1</u>. *Let* $\phi(\tau)$ *be a family of connections in* P *depending on a parameter* τ . *Let* $\psi(\tau) = \partial\phi/\partial\tau$ *and*

$$(87) \qquad V(\tau) = \int_0^1 t^{h-1}F(\psi,\phi(\tau),d\phi(\tau) - \tfrac{1}{2}t[\phi(\tau),\phi(\tau)], \ldots, d\phi(\tau) - \tfrac{1}{2}t[\phi(\tau),\phi(\tau)])dt$$

Then

$$(88) \qquad \frac{\partial}{\partial\tau} TF(\Phi(\tau)) = h(h-1)dV(\tau) + hF(\psi,\Phi(\tau),\ldots,\Phi(\tau)) .$$

To avoid long expressions we will use the *convention* that if F contains fewer than h arguments the last one is to be repeated a number of times so as to make F a function of h arguments. Thus

$$(89) \qquad F(X) = F(X,\ldots,X), F(X,Y) = F(X,\underbrace{Y,\ldots,Y}_{h-1}), \text{ etc.}$$

With this convention in mind we find

$$dF(\psi,\phi(\tau),d\phi(\tau) - \tfrac{1}{2}t[\phi(\tau),\phi(\tau)]) =$$

$$F(d\psi,\phi(\tau),d\phi(\tau) - \tfrac{1}{2}t[\phi(\tau),\phi(\tau)]) - F(\psi,d\phi(\tau),d\phi(\tau) - \tfrac{1}{2}t[\phi(\tau),\phi(\tau)])$$

$$- (h-2)tF(\psi,\phi(\tau),[d\phi(\tau) - \tfrac{1}{2}t[\phi(\tau),\phi(\tau)],\phi(\tau)],d\phi(\tau) - \tfrac{1}{2}t[\phi(\tau),\phi(\tau)]) .$$

By (66) the last term is equal to

$$-tF([\phi(\tau),\psi],\phi(\tau),d\phi(\tau) - \tfrac{1}{2}t[\phi(\tau),\phi(\tau)])$$

$$+ tF(\psi,[\phi(\tau),\phi(\tau)],d\phi(\tau) - \tfrac{1}{2}t[\phi(\tau),\phi(\tau)]) .$$

We have therefore

$$dF(\psi,\phi(\tau),d\phi(\tau) - \frac{1}{2} t[\phi(\tau),\phi(\tau)])$$

$$= F(d\psi - t([\phi(\tau),\psi],\phi(\tau),d\phi(\tau) - \frac{1}{2} t[\phi(\tau),\phi(\tau)])$$

$$- F(\psi,d\phi(\tau)-t[\phi(\tau),\phi(\tau)],d\phi(\tau) - \frac{1}{2} t[\phi(\tau),\phi(\tau)]) .$$

On the other hand, we have

$$h^{-1} \frac{\partial}{\partial\tau} TF(\phi(\tau))$$

$$= \int_0^1 t^{h-1}F(\psi,d\phi(\tau) - \frac{1}{2} t[\phi(\tau),\phi(\tau)])dt$$

$$+ (h-1) \int_0^1 t^{h-1}F(\phi(\tau),d\psi - t[\phi(\tau),\psi],d\phi(\tau) - \frac{1}{2} t[\phi(\tau),\phi(\tau)])dt .$$

It follows that

$$h^{-1} \frac{\partial}{\partial\tau} TF(\phi(\tau)) - (h-1)dV(\tau)$$

$$= h \int_0^1 t^{h-1}F\left(\psi,d\phi(\tau) - \frac{2h-1}{2h} t[\phi(\tau),\phi(\tau)],d\phi(\tau) - \frac{1}{2} t[\phi(\tau),\phi(\tau)]\right) dt .$$

To simplify the last integral we introduce the curvature form

$$\Phi(\tau) = d\phi(\tau) - \frac{1}{2} [\phi(\tau),\phi(\tau)] \tag{90}$$

of the connection $\phi(\tau)$. Putting

$$a = \frac{2h-1}{h} ,$$

the integrand above, up to the factor t^{h-1} , can be expanded:

$$F(\psi,d\phi(\tau) - \frac{1}{2}\, at[\phi(\tau),\phi(\tau)],d\phi(\tau) - \frac{1}{2}\, t[\phi(\tau),\phi(\tau)])$$

$$= F(\psi,\Phi(\tau) + \frac{1}{2}\,(1-at)[\phi(\tau),\phi(\tau)],\Phi(\tau) + \frac{1}{2}\,(1-t)[\phi(\tau),\phi(\tau)])$$

$$= F(\psi,\Phi(\tau))$$

$$+ \sum_{1\leqq r\leqq h-1} c_r(t)F(\psi,\underbrace{\Phi(\tau),\ldots,\Phi(\tau)}_{h-1-r}\,,\ \underbrace{\frac{1}{2}\,[\phi(\tau),\phi(\tau)],\ldots,\ \frac{1}{2}\,[\phi(\tau),\phi(\tau)]}_{r})$$

where

$$c_r(t) \quad = \quad \binom{h-2}{r}(1-t)^r + \binom{h-2}{r-1}(1-t)^{r-1}(1-at) \ .$$

From elementary calculus we know

$$\int_0^1 t^m(1-t)^n dt = \frac{m!n!}{(m+n+1)!}\ , \qquad m \geqq 0\ ,\ n \geqq 0\ .$$

Using this it is immediately verified that

$$\int_0^1 t^{h-1}c_r(t)dt = 0\ , \qquad 1 \leqq r \leqq h - 1\ .$$

This proves the lemma.

From the lemma follows immediately the theorem:

Theorem 4.1. *Let* $\pi\colon P \to M$ *be a principal bundle with the group* G *, and let* $F \in I(G)$ *be an invariant polynomial. Let* $\phi(\tau)$ *be a family of connnections, with the curvature form* $\Phi(\tau)$ *, which satisfy the conditions*

$$\text{(91)} \qquad F(\partial\phi(\tau)/\partial\tau,\Phi(\tau),\ldots,\Phi(\tau)) = 0\ ,$$
$$F(\Phi(\tau),\ldots,\Phi(\tau)) = 0\ .$$

Then the cohomology class $\{TF(\phi(\tau))\}$ *is independent of* τ .

As h is the degree of F , the conditions (91) are automatically satisfied when $2h > \dim M + 1$.

Equivalently the conditions (91) can be written in terms of a local chart. By (46) and using the fact that F is an invariant polynomial, we can write (91) as

(92)
$$F(\partial\Theta_U(\tau)/\partial\tau, \Theta_U(\tau), \ldots, \Theta_U(\tau)) = 0 ,$$
$$F(\Theta_U(\tau), \ldots, \Theta_U(\tau)) = 0 .$$

We now apply these results to the principal bundle of the tangent bundle of a manifold M of dimension n, so that the structure group is $G = GL(n;R)$. The connection will be the Levi-Civita connection of a riemannian metric in M and it will have special properties. We use a local chart with the coordinates x^i, $1 \leq i,j,k, \ell \leq n$, and we will omit the subscript U in our notations. The riemannian metric is given by the scalar products

(93)
$$h_{ij} = h_{ji} = \left(\frac{\partial}{\partial x^i}, \frac{\partial}{\partial x^j}\right) ,$$

which are the elements of a positive definite symmetric matrix:

(94)
$$H = {}^tH = (h_{ij}) > 0 .$$

Let the connection matrix be

(95)
$$\theta = (\theta_i^j) , \quad \theta_i^j = \sum_k \Gamma_{ik}^j \, dx^k .$$

It is determined by the conditions

(96)
$$dH - \theta H - H{}^t\theta = 0 , \quad \Gamma_{ik}^j = \Gamma_{ki}^j .$$

The curvature form is given by

(97)
$$\Theta = d\theta - \theta \wedge \theta .$$

Exterior differentiation of (96) gives

(98)
$$\Theta H + H{}^t\Theta = 0 ,$$

i.e., the matrix ΘH is anti-symmetric.

Lemma 4.2. *Let* F *be an invariant polynomial of odd degree* h *, and* Θ *the curvature matrix of the Levi-Civita connection of a riemannian metric. Then*

$$F(\Theta) = 0 . \tag{99}$$

To prove this, notice that F clearly has the property:

$$F(\Theta) = F({}^t\Theta) . \tag{100}$$

By (98) this is equal to

$$F(-H^{-1}\Theta H) = (-1)^h F(\Theta) .$$

Hence we have (99) when h is odd.

We write

$$\Theta = (\Theta_i^j) , \tag{101}$$

where we set

$$\Theta_i^j = -\frac{1}{2}\sum_{k,\ell} R_{ik\ell}^j dx^k \wedge dx^\ell , \quad R_{ik\ell}^j + R_{i\ell k}^j = 0 . \tag{102}$$

The $R_{ik\ell}^j$ define the Riemann curvature tensor and satisfy the symmetry relations

$$\begin{gathered} R_{ik\ell}^j + R_{i\ell k}^j = 0 , \\ R_{ik\ell}^j + R_{k\ell i}^j + R_{\ell ik}^j = 0 . \end{gathered} \tag{103}$$

The last relation can also be written

$$\sum_i dx^i \wedge \Theta_i^j = 0 . \tag{104}$$

We will consider the case of a conformal family of riemannian metrics, given by the matrix $H(\tau) = \exp(2\sigma\tau)H$, where σ is a scalar function and τ is the parameter. Then we have ([20], p. 89) the matrix equation

(105) $$\frac{1}{\tau}(\theta(\tau) - \theta(0)) = d\sigma I + \alpha + \beta ,$$

where I is the unit matrix and

(106) $$\alpha = \left(\frac{\partial\sigma}{\partial x^i} dx^j\right) , \quad \beta = -\left(\sum_k h_{ik} dx^k \sum_\ell \frac{\partial\sigma}{\partial x^\ell} h^{j\ell}\right)$$
$$H^{-1} = (h^{ij}) .$$

Lemma 4.3. *Let* F *be an invariant polynomial of even degree* $2s$. *For the Levi-Civita connections of a conformal family of riemannian metrics we have*

(107) $$F\left(\frac{\partial\theta(\tau)}{\partial\tau} , \theta(\tau)\right) = 0 .$$

In fact, the left-hand side of this equation is equal to

$$d\sigma F(I,\Theta(\tau)) + F(\alpha,\Theta(\tau)) + F(\beta,\Theta(\tau)) .$$

By Lemma 4.2 the first term is zero. By the fundamental theorem on vector invariants every term in $F(\alpha,\theta(\tau))$ contains a factor $\Sigma dx^j \wedge \Theta^i_j(\tau)$, which is zero by (104). Similarly, every term in $F(\beta,\Theta(\tau))$ contains a factor $\Sigma_{i,k} h_{ik} dx^k \wedge \Theta^i_j(\tau)$, which is also seen to be zero. Thus the lemma is proved.

This lemma, together with the formula (88), gives the theorem:

Theorem 4.2. *Let* $\pi\colon P \to M$ *be the principal bundle of the tangent bundle of a manifold* M *of dimension* n . *Let* $F \in I(GL(n;R))$ *be an invariant polynomial of degree* $2s$. *Let* ϕ *and* ϕ^* *be the connection forms of the Levi-Civita connections of two riemannian metrics on* M , *which are conformal to each other. Then there exists a form* W *of degree* $4s-2$ *in* P , *such that*

(108) $$TF(\phi^*) - TF(\phi) = dW .$$

Corollary 4.1. *The form* $F(\Phi)$ *remains invariant under a conformal transformation of the riemannian metric.*

More precisely, A. Avez ([3]) expressed $F(\Phi)$ in terms of the Weyl conformal tensor of the riemannian metric.

Corollary 4.2. *If* $F(\Phi) = 0$, *then* $TF(\phi)$ *defines an element* $\{TF(\phi)\} \in H^{4s-1}(P,R)$, *which depends only on the underlying conformal structure of the riemannian manifold* M .

In exactly the same way one can establish results concerning a projective transformation of the riemannian metric, i.e., a change of the riemannian metric which leaves the geodesics invariant. Such a change is described by a form $\lambda = \Sigma_i a_i(x)dx^i$ and the connection forms are related by ([20], p. 132)

(109) $$\theta^* - \theta = \lambda I + \alpha ,$$

where

(110) $$\alpha = (a_i dx^j) .$$

The connections θ and θ^* can be joined by the family $\theta(\tau)$, $0 \leqq \tau \leqq 1$, given by

(111) $$\frac{1}{\tau}(\theta(\tau) - \theta) = \lambda I + \alpha .$$

The above arguments apply and we conclude that $F(\Phi)$ is a projective invariant and that, if $F(\Phi) = 0$, the cohomology class $\{TF(\phi)\}$ in P is also a projective invariant.

In order to utilize our secondary invariants we look for cases where $F(\Phi) = 0$. One is the situation which occurs in Bott's theorem on foliations discussed above. Recently this has given rise to an active development in the works of Bott, Haefliger, etc. [35].

Another case concerns with immersed submanifolds M in the euclidean space E^N of dimension $N = n + h$, which we will discuss in some detail. The basic fact is the commutative diagram

$$(112)\qquad \begin{array}{ccc} P & \xrightarrow{\tilde{g}} & O(n+h)/O(h) \\ \pi\downarrow & & \downarrow\pi_0 \\ M & \xrightarrow{g} & O(n+h)/\{O(n)\times O(h)\}\ . \end{array}$$

Here P is the bundle of orthonormal frames over M and g and $\tilde{g}$ are the Gauss mappings defined by parallelisms in the ambient Euclidean space. The bundle at the right-hand side of the diagram has a canonical connection described at the end of §2. We denote its connection and curvature forms by $\tilde{\phi}$ and $\tilde{\Phi}$ respectively; they are therefore anti-symmetric matrices of forms. Then the Levi-Civita connection is given by $\phi = \tilde{g}^*\tilde{\phi}$ and its curvature form is $\Phi = \tilde{g}^*\tilde{\Phi}$. In fact, this was the original definition of Levi-Civita of his connection, generalizing a classical construction for surfaces in E^3 .

We put

$$(113)\qquad P_k(\Phi) = E_{2k}(\Phi)\ ,$$

where the latter are defined in (84); these will be called the *Pontrjagin forms*. The dual Pontrjagin forms are introduced by the equation

$$(114)\qquad \sum_{k\geqq 0} P_k(\Phi) \sum_{j\geqq 0} P_j^{\perp}(\Phi) = 1\ , \qquad P_0 = P_0^{\perp} = 1\ ,$$

and are uniquely determined. By the duality theorem on Pontrjagin classes, the cohomology classes $\{P_j^{\perp}(\Phi)\} \in H^{4j}(M,R)$ are the Pontrjagin classes of the normal bundle of M in E^N . Since the normal bundle has fiber dimension h , we have

$$(115)\qquad \{P_j^{\perp}(\Phi)\} = 0\ , \qquad \left[\frac{h}{2}\right] + 1 \leqq j\ .$$

We will show that the forms $P_j^{\perp}(\Phi)$ themselves are zero. In fact, we have

$$\{P_j^{\perp}(\tilde{\Phi})\} = 0\ . \qquad \left[\frac{h}{2}\right] + 1 \leqq j\ .$$

But the Grassmann manifold is a symmetric riemannian manifold and the form $P_j^{\perp}(\tilde{\Phi})$ is invariant under the action of the group $O(n+h)$. Hence, in the range of j described above, we have $P_j^{\perp}(\tilde{\Phi}) = 0$ and therefore

$$P_j^{\perp}(\Phi) = \tilde{g}^* P_j(\tilde{\Phi}) = 0 .$$

It is thus possible to apply the construction of secondary invariants to the invariant polynomials $P_j^{\perp}$. We will state our general theorem as follows:

Theorem 4.3. *Let* M *be a compact manifold of dimension* n, *with a riemannian metric* ds^2. *Necessary conditions for its conformal immersion in* E^{n+h} *are:*

$$(116) \qquad P_j^{\perp}(ds^2) = 0 , \quad \left[\frac{h}{2}\right] + 1 \leqq j ,$$

$$(117) \qquad \{\tfrac{1}{2} TP_j^{\perp}(ds^2)\} \in H^{4j-1}(P,Z) , \quad \left[\frac{h}{2}\right] + 1 \leqq j \leqq \left[\frac{n-1}{2}\right] ,$$

where we use the argument ds^2 *to replace its Levi-Civita connection in the notation.*

Conditions (116) follows from the above discussions. The proof of (117) is lengthy and can be found in [19].

We will carry out our construction for $P_1 = -P_1^{\perp}$. By (113) and (84) we have

$$(118) \qquad P_1(\Phi) = \frac{1}{8\pi^2} \sum_{i,j} (\Phi_i^i \Phi_j^j - \Phi_i^j \Phi_j^i) .$$

In the notation of (72) we introduce the form

$$(119) \qquad \Phi_t = t(d\phi - t\phi \wedge \phi) = t\{\Phi + (1-t)\phi \wedge \phi\} ,$$

so that we find the polarized form

(120)

$$P_1(\phi,\Phi_t) = \frac{t}{8\pi^2} \sum_{i,j,k} \{\phi^i_i \wedge \Phi^j_j - \phi^j_i \wedge \Phi^i_j - (1-t)\phi^j_i \wedge \phi^k_j \wedge \phi^i_k\} \ .$$

It follows that

$$\text{(121)} \quad TP_1(\phi) = 2 \int_0^1 P_1(\phi,\Phi_t)dt$$

$$= \frac{1}{8\pi^2} \sum_{i,j,k} \{\phi^i_i \wedge \Phi^j_j - \phi^j_i \wedge \Phi^i_j - \frac{1}{3}\phi^j_i \wedge \phi^k_j \wedge \phi^i_k\} \ .$$

When restricted to orthonormal frames the matrices

$$\phi = (\phi_{ij}) \ , \qquad \Phi = (\Phi_{ij})$$

are anti-symmetric and (121) simplifies to

$$\text{(122)} \quad TP_1(\phi) = \frac{1}{8\pi^2} \left\{ \sum_{i,j} \phi_{ij} \wedge \Phi_{ij} - \frac{1}{3} \sum_{i,j,k} \phi_{ij} \wedge \phi_{jk} \wedge \phi_{ki} \right\}$$

When M is of dimension 3, $P_1(\Phi)$ vanishes for dimension reasons and we get a closed form $TP_1(\phi)$ in the bundle P of orthonormal frames. In view of Theorem 4.3 we write

(123)

$$\frac{1}{2} TP_1(\phi) = \frac{1}{8\pi^2} \sum_{1 \leqq i<j \leqq 3} \phi_{ij} \wedge \Phi_{ij} - \frac{1}{8\pi^2} \phi_{12} \wedge \phi_{23} \wedge \phi_{31} \ ,$$

and we find

$$\text{(124)} \quad \int_{\pi^{-1}(x)} \frac{1}{2} TP_1(\phi) = 1 \ ,$$

when the fibers $\pi^{-1}(x)$, $x \in M$, are properly oriented.

Suppose M be compact and orientable. Our form $\frac{1}{2} TP_1(\phi)$ gives rise to an invariant $J(\phi) = J(ds^2) \in R/Z$ as follows: It is known that M is parallelizable, so that a section $s\colon M \to P$ exists. The integral

$$(125) \qquad I(s) = \int_{sM} \tfrac{1}{2} TP_1(\phi)$$

is a real number. For another section $s': M \to P$ the difference $I(s) - I(s')$ is an integer, since P is homologically equivalent to the product $M \times \pi^{-1}(x)$ modulo torsion and $\frac{1}{2} TP_1(\phi)$ satisfies (124). The invariant $J(ds^2)$ is defined to be $I(s)$ mod 1. By Corollary 4.2 it depends only on the conformal structure on M and by Theorem 4.3 it is zero if M can be conformally immersed in E^4.

To show that our invariants are not vacuous we wish to calculate $J(ds^2)$ for $M = SO(3)$ with its biinvariant riemannian metric. M is therefore the elliptic space in non-euclidean geometry. Let $\omega_{ij} = -\omega_{ji}$, $1 \leqq i, j \leqq 3$, be the Maurer-Cartan forms in $SO(3)$, so that the structure equations are

$$(126) \qquad d\omega_{ik} = \sum_j \omega_{ij} \wedge \omega_{jk}, \qquad 1 \leqq i,j,k \leqq 3 .$$

Its biinvariant metric is given by

$$(127) \qquad ds^2 = \omega_{12}^2 + \omega_{13}^2 + \omega_{23}^2 .$$

In writing these equations we have chosen a basis in the Lie algebra of $SO(3)$ and hence, by right translations, a frame field in the manifold $SO(3)$. It will be convenient to choose our notation so that the equations remain invariant under a cyclic permutation of 1, 2, 3. We set

$$(128) \qquad \alpha_i = \omega_{jk}, \quad i,j,k = \text{cyclic permutation of } 1, 2, 3 .$$

Then (127) becomes

$$(129) \qquad ds^2 = \alpha_1^2 + \alpha_2^2 + \alpha_3^2 .$$

The connection and curvature forms

$$\theta_{ij} = -\theta_{ji}, \qquad \Theta_{ij} = -\Theta_{ji}$$

are determined by the equations

(130)
$$d\alpha_i = \sum_j \alpha_j \wedge \theta_{ji} ,$$
$$d\theta_{ik} - \sum_j \theta_{ij} \wedge \theta_{jk} = \Theta_{ik} .$$

Comparing these with the structure equations (126), we find

(131) $$\theta_{ij} = \frac{1}{2}\alpha_k , \qquad \Theta_{ij} = -\frac{1}{4}\alpha_i \wedge \alpha_j .$$

It follows that

(132) $$\frac{1}{2} TP_1(\phi) = -\frac{1}{16\pi^2}\alpha_1 \wedge \alpha_2 \wedge \alpha_3 .$$

Since the total volume of $SO(3)$ is $8\pi^2$, we get $J = \frac{1}{2}$ for $M = SO(3)$ with the biinvariant metric. It is to be observed that J remains unchanged when the metric is modified by a constant positive factor because it is a conformal invariant. As a consequence we have the theorem: *The non-Euclidean elliptic space cannot be conformally immersed in* E^4 .

This is a global theorem, because the space is isometrically covered by the three-dimensional sphere of constant curvature and can certainly be locally isometrically imbedded in E^4 . On the other hand, by a theorem of M. Hirsch it can be globally differentiably immersed in E^4 .

Remark. The cohomology classes $\{TF(\phi)\}$ with real coefficients, when they are defined, are in the principal bundle P . It is possible, using the connection, to define cohomology classes with coefficients R/Z in the base manifold. These invariants are called *Simons characters* (unpublished).

5. Vector Fields and Characteristic Numbers

We will give an account of results of Bott, Baum, and Cheeger on relations between the characteristic numbers of a manifold and the behavior at the zeroes of a vector field which satisfies certain conditions. As noted by these authors, the Weil homomorphism plays a fundamental rôle in these results.

We will deal with the tangent bundle of a real or complex manifold, so that the structure group G is the real or complex linear group and is, in the case of a riemannian manifold, the orthogonal group. As in previous sections we consider these groups as matrix groups and their Lie algebras as spaces of matrices. Adjoint action is given by

$$\mathrm{ad}(A)X = AXA^{-1}, \quad A \in G, \quad X \in g.$$

An h-linear function F is invariant if

$$F(AX_1A^{-1},\ldots,AX_hA^{-1}) = F(X_1,\ldots,X_h), \tag{133}$$
$$X_i \in g, \quad \text{all} \quad A \in G.$$

Consider first the case of a complex hermitian manifold M of complex dimension m. This means that in the complex tangent spaces T_x, $x \in M$, of M there is given a C^∞-family of positive definite hermitian scalar products $H(\xi,\eta)$, ξ, $\eta \in T_x$, which is linear in ξ and antilinear in η. In local coordinates z^i, $1 \leqq i,j,k,\ell \leqq m$, the hermitian structure is defined by the scalar products of the basis vectors:

$$h_{ik} = H\left(\frac{\partial}{\partial z^i}, \frac{\partial}{\partial z^k}\right) = \bar{h}_{ki}, \tag{134}$$

and the matrix

$$H = {}^t\bar{H} = (h_{ik})$$

is positive definite. A complex vector field is given by

$$\xi = \sum_i \xi^i \frac{\partial}{\partial z^i}. \tag{135}$$

It is called holomorphic, if the components ξ^i are holomorphic functions in z^k.

A connection

(136) $$D\left(\frac{\partial}{\partial z^i}\right) = \sum_k \omega_i^k \frac{\partial}{\partial z^k} .$$

is uniquely determined by the conditions:

1. For two holomorphic vector fields ξ, η, defined locally,

(137) $$dH(\xi,\eta) = H(D\xi,\eta) + H(\xi,D\eta) ;$$

2. The connection forms ω_i^k are of bidegree (1,0). In fact, the first condition can be written as

(138) $$dh_{ik} = \sum_j \omega_i^j h_{jk} + \sum_j h_{ij}\bar{\omega}_k^j ,$$

and the second condition gives

(139) $$\partial h_{ik} = \sum_j \omega_i^j h_{jk} ,$$

which completely determines the connection.

The equations can be shortened by the introduction of the matrix

(140) $$\omega = (\omega_i^j)$$

Then (139) can be written

(139a) $$\omega = \partial H \cdot H^{-1} .$$

On exterior differentiation we get

(141) $$\partial\omega - \omega \wedge \omega = 0 .$$

On the other hand, the curvature matrix is defined by

(142) $$\Omega = d\omega - \omega \wedge \omega .$$

Using (141) we get

(143) $$\Omega = \bar{\partial}\omega ,$$

so that it is of bidegree (1,1).

It is important to study the effect on these matrices under a change of chart. If the new coordinates are z^{*i} and if the new quantities are denoted by the same notations with asterisks, we have the easily verified equations

(144) $$\begin{aligned} H^* &= JH{}^t\bar{J} \\ \omega^* &= \partial J J^{-1} + J\omega J^{-1} , \\ \Omega^* &= J\Omega J^{-1} , \end{aligned}$$

where

(145) $$J = \left(\frac{\partial z^j}{\partial z^{*i}}\right)$$

is the Jacobian matrix. In particular, Ω is an endomorphism-valued two-form or, what is the same, a two-form with values in $TM \otimes T^*M$.

Let

(146) $$\omega_i^j = \sum_k \Gamma_{ik}^j \, dz^k .$$

Then

(147) $$T_{ik}^j = \Gamma_{ik}^j - \Gamma_{ki}^j$$

are the components of a section of the bundle $TM \otimes T^*M \otimes T^*M$. It defines the so-called torsion tensor.

Let ξ be a holomorphic vector field. Then

(148) $$D\xi = \sum_{i,j} \xi^i{}_{,j} dz^j \otimes \frac{\partial}{\partial z^i}$$

is an element of $\Gamma(T^*M \otimes TM)$ and is a field of endomorphisms. If we put

(149) $$\Xi = (\xi^i{}_{,j})$$

we have

(150) $$\Xi^* = J \Xi J^{-1} .$$

Observe that at $\xi = 0$, Ξ is the matrix of the partial derivatives of ξ^i . A zero of ξ is called *nondegenerate* if $\det\Xi \neq 0$.

Another field of endomorphisms is given by the components $\Sigma_k T^j_{ik} \xi^k$. Combining the two, we have the field of endomorphisms Ξ_λ with the components $\xi^j_{,i} + \lambda\Sigma_k T^j_{ik}\xi^k$, having λ as a parameter. Clearly under a change of chart we have

(151) $$\Xi^*_\lambda = J \Xi_\lambda J^{-1} .$$

Using the local chart we immediately verify that

(152) $$\bar{\partial}\Xi_1 = - i(\xi)\bar{\partial}\omega ,$$

where $i(\xi)$ denotes the interior product by the vector ξ . It follows that

$$i(\xi)\Omega = i(\xi)\bar{\partial}\omega = - \bar{\partial}\Xi_1 .$$

By putting $E = - \Xi_1$, we have

(153) $$\bar{\partial}E = i(\xi)\Omega .$$

Perhaps the simplest result after the Hopf formula (2) on relations between characteristic numbers and vector fields is the following theorem of Bott.

Theorem 5.1. *Let* M *be a compact complex hermitian manifold of complex dimension* m , *whose curvature matrix is* Ω . *Let* F *be an invariant polynomial of degree* m *relative to the group* GL(m;C). *Suppose* ξ *be a holomorphic vector field on* M *with nondegenerate isolated zeroes. Then*

(154) $$\left(\frac{i}{2\pi}\right)^m \int_M F(\Omega) = \sum_{\text{zero of } \xi} F(\Xi)/\det\Xi .$$

We will sketch a proof of this theorem, whose idea is quite simple. It is to write $F(\Omega)$ as a derived form in M-(zero of ξ) and to apply Stokes' Theorem. We polarize F and insert as arguments both Ω and E, which are both endomorphism-valued, i.e., we put

$$F^{(r)}(\Omega) = \binom{m}{r} F(\underbrace{E,\dots,E}_{r},\underbrace{\Omega,\dots,\Omega}_{m-r}), \quad 0 \leqq r \leqq m, \tag{155}$$

so that $F^{(r)}(\Omega)$ is a form of bidegree $(m-r,\ m-r)$ in M. Using (153) we immediately get

$$i(\xi)F^{(r)}(\Omega) = \bar\partial F^{(r+1)}(\Omega), \quad 0 \leqq r \leqq m-1. \tag{156}$$

The vector field ξ gives rise to the one-form

$$\pi = \sum_{i,k} h_{ik} dz^i \bar\xi^k \wedge \sum_{i,k} h_{ik} \xi^i \bar\xi^k, \quad \xi \neq 0, \tag{157}$$

satisfying $i(\xi)\pi = 1$. It is easily verified that

$$i(\xi)\bar\partial\pi = 0. \tag{158}$$

Since both $i(\xi)$ and $\bar\partial$ are anti-derivations, we find

$$i(\xi)\bar\partial(\pi \wedge (\bar\partial\pi)^{r-1} \wedge F^{(r)}(\Omega)) =$$

$$(\bar\partial\pi)^r i(\xi)F^{(r)}(\Omega) - (\bar\partial\pi)^{r-1} i(\xi)F^{(r-1)}(\Omega), \quad 1 \leqq r \leqq m,$$

which gives

$$i(\xi)\left\{\sum_{1\leqq r\leqq m} \bar\partial(\pi \wedge (\bar\partial\pi)^{r-1} \wedge F^{(r)}(\Omega)) + F(\Omega)\right\} = 0.$$

The form in the braces is of bidegree (m,m). Since we have assumed $\xi \neq 0$, it follows that

$$\bar\partial\left(\sum_{1\leqq r\leqq m} \pi \wedge (\bar\partial\pi)^{r-1} \wedge F^{(r)}(\Omega)\right) + F(\Omega) = 0.$$

By consideration of the bidegree the operator $\bar\partial$ at the left can be replaced by d. This gives

(159) $$F(\Omega) = d\Pi \,, \qquad \text{in } M\text{-(zero of } \xi) \,,$$

where

(160) $$\Pi = - \sum_{1 \leq r \leq m} \pi \wedge (\bar{\partial}\pi)^{r-1} \wedge F^{(r)}(\Omega) \,.$$

The formula (159) localizes the problem of integrating $F(\Omega)$.

To evaluate the integral of $F(\Omega)$ over M it suffices to integrate Π over the spheres S_ε of radius ε about the zeroes of ξ and take the limit of the integral as $\varepsilon \to 0$. Since the integral is a characteristic number, its value is independent of the choice of an hermitian metric on M. We choose the latter so that it has a simple behavior at the zeroes of ξ. In fact, let z^i be a local coordinate system centered at an isolated zero of ξ. Let λ_i be the eigenvalues of the matrix Ξ at the origin. The nondegeneracy of the zero is equivalent to the condition $\lambda_1 \ldots \lambda_m \neq 0$. We can choose the coordinates z^i so that in a sufficiently small neighborhood we have

(161) $$\xi = \sum_i \lambda_i z^i \frac{\partial}{\partial z^i} \,.$$

Suppose the metric be

(162) $$ds^2 = \sum_i \frac{1}{|\lambda_i|^2} dz^i \, d\bar{z}^i \,.$$

Then we have

(163) $$\pi = \sum_i \lambda_i^{-1} \bar{z}^i dz^i \Big/ \sum_i |z^i|^2$$

and

$$(\Sigma |z^i|^2) \, \bar{\partial}\pi = - \sum_i \lambda_i^{-1} \, dz^i \wedge d\bar{z}^i + \pi \wedge (\ldots) \,,$$

$$(\Sigma |z^i|^2)^m \pi \wedge (\bar{\partial}\pi)^{m-1}$$

$$= \pm \frac{(m-1)!}{\lambda_1 \ldots \lambda_m} \left\{ \sum_i dz^1 \wedge d\bar{z}^1 \wedge \ldots \wedge \bar{z}^i dz^i \wedge \ldots \wedge dz^m \wedge d\bar{z}^m \right\}$$

The expression between the braces is a multiple of the volume element (relative to the metric $\Sigma_i\, dz^i\, d\bar{z}^i$) of $S_\varepsilon = \{z|\Sigma|z^i|^2 = \varepsilon^2\}$, when restricted to S_ε. For the metric (162) we have clearly $\Omega = 0$, so that

$$\Pi = -\,\pi \wedge (\bar{\partial}\pi)^{m-1} F(E,\ldots,E)\ . \tag{164}$$

As $\varepsilon \to 0$, we obtain

$$\int_M F\left(\frac{i}{2\pi}\,\Omega\right) = C \sum_{\text{zero of }\xi} F(\Xi)/\det \Xi\ ,$$

where C is a universal constant independent of F. By putting $F = \det$, we find $C = 1$. This proves the theorem.

We can formulate Theorem 5.1 in terms of the Chern classes of M. In fact, let F_k, $1 \leqq k \leqq m$, be the functions introduced in (82). There exists to F a polynomial $\tilde{F}$ such that

$$F\left(\frac{i}{2\pi}\,\Omega\right) = \tilde{F}(F_1(\Omega),\ldots,F_m(\Omega))\ . \tag{165}$$

Then

$$\int_M F\left(\frac{i}{2\pi}\,\Omega\right) = \int_M \tilde{F}(c_1(M),\ldots,c_m(M))\ . \tag{166}$$

On the other hand, each summand at the right-hand side of (154) can be written

$$\tilde{F}(\sigma_1,\ldots,\sigma_m)/\sigma_m\ ,$$

where σ_i, $1 \leqq i \leqq m$, is the ith elementary symmetric function of the eigenvalues of Ξ. Theorem 5.1 can be stated as follows:

Theorem 5.2. *Let* M *be a compact complex manifold of complex dimension* m. *Let*

$$c^\alpha(M) = c_1^{\alpha_1}(M)\ldots c_m^{\alpha_m}(M), \qquad \alpha_1 + 2\alpha_2 + \ldots + m\alpha_m = m\ . \tag{167}$$

Suppose ξ *be a holomorphic vector field with non-degenerate isolated zeroes. Then*

(168) $$\int_M c^{\alpha}(M) = \sum_{\text{zero of } \xi} \sigma_1^{\alpha_1} \cdots \sigma_m^{\alpha_m}/\sigma_m ,$$

where σ_i , $1 \leq i \leq m$, *is the ith elementary symmetric function of the eigenvalues of the matrix* $\Xi = (\partial\xi^i/\partial z^k)$ *at* $\xi = 0$.

Baum and Bott extended Theorem 5.1 to meromorphic vector fields with isolated zeroes ([5]). They used an algebraic geometrical method (cf. also new differential geometrical proof by Chern [34]).

The theorem has a real analogue treated by Bott and further pursued by Baum and Cheeger ([6], [18]). It concerns with the Killing vector fields of a compact oriented riemannian manifold of even dimension 2m. The Killing equations are classically written ([20])

(169) $$\xi_{i,j} + \xi_{j,i} = 0 ,$$

so that the matrix $\Xi = (\xi_{i,j})$ is anti-symmetric. Its eigenvalues are of the form $\pm i\lambda_j$, $1 \leq j \leq m$, where λ_j is real. We denote by σ_j the jth elementary symmetric function of $\lambda_1^2, \ldots, \lambda_m^2$ and let $\tau = \lambda_1 \cdots \lambda_m$. Then $\tau = \sigma_m^{1/2}$ depends on the orientation of M and changes its sign under a reversal of the orientation. A zero of ξ is called *non-degenerate* if $\tau \neq 0$. We consider an invariant polynomial F of degree m with respect to the group SO(2m) , so that its arguments are (2m × 2m) anti-symmetric matrices. Into F we substitute the curvature forms of the riemannian metric. The study of its integral leads to the theorem:

<u>Theorem 5.3</u>. *Let* M *be a compact oriented riemannian manifold of dimension* 2m . *Let* $p_i(M)$, $1 \leq i \leq [m/2]$, *be the Pontrjagin classes of* M *and* e(M) *be its Euler class. Let* ξ *be a Killing vector field with nondegenerate isolated zeroes. Then*

$$(170)\qquad \int_M p_1^{\alpha_1}\cdots p_h^{\alpha_h} e(M)^\beta = \sum_{\text{zero of } \xi} \sigma_1^{\alpha_1}\cdots\sigma_h^{\alpha_h}\tau^{\beta-1}\ , \quad h = \left[\frac{m}{2}\right]\ ,$$

where

$$2(\alpha_1 + 2\alpha_2 + \ldots + h\alpha_h) + \beta m = m\ .$$

If the vector field generates a compact group, such results have an alternative treatment by the Atiyah-Singer G-index theory. Cf. [2].

6. Holomorphic Curves

We have given in the above several applications of the Weil homomorphism, i.e., the representation of characteristic classes by curvature forms. Perhaps the most important ones remain to come from the study of noncompact manifolds, which is a far more difficult subject than the case of compact manifolds. For non-compact manifolds standard approaches to characteristic classes (such as an axiomatic treatment) do not apply and the curvature representation plays a more vital rôle.

To get deep results it is probably necessary to impose on the problems conditions in the form of differential equations or differential inequalities. An example of such conditions is the Cauchy-Riemann equations in complex function theory; perhaps no other differential system has been as thoroughly studied. In this section we will consider non-compact holomorphic curves in the complex projective space and show that the curvature forms of hermitian line bundles over a curve play a fundamental rôle in the theory of value distributions of Nevanlinna-Weyl-Ahlfors. It is to be pointed out that a non-compact holomorphic curve in the complex projective line is exactly what is called a meromorphic function, which is generally non-algebraic. Thus the results do have wider scope than the compact holomorphic curves.

Let M be a complex manifold of complex dimension m and let

$\pi: E \to M$ be a holomorphic line bundle. This means that M has an open covering $\{U, V, \ldots\}$ such that to each U there is a chart $\psi_U: \pi^{-1}(U) \to U \times C$, with $\psi_U(z) = (\pi(z) = x, y_U(z))$, $z \in \pi^{-1}(U)$; the local charts are related by the equation

$$y_U g_{UV} = y_V \,, \qquad \text{in } U \cap V \neq \emptyset \,, \tag{171}$$

where $g_{UV}: U \cap V \to C - \{0\}$ is holomorphic.

The holomorphy of the transition functions g_{UV} has important implications. Let an hermitian norm be given in E, i.e., a C^∞-function $h_U > 0$ in each U, such that

$$\|y\|^2 = h_U^{-1}|y_U|^2 = h_V^{-1}|y_V|^2 \qquad \text{in } U \cap V \,. \tag{172}$$

Equation (172) is equivalent to

$$h_U|g_{UV}|^2 = h_V \tag{172a}$$

It follows that

$$\Omega = \frac{i}{2\pi}\,\partial\bar{\partial}\log h_U \tag{173}$$

is independent of U. Ω defines a closed form of bidegree (1,1) in M, the curvature form of the hermitian line bundle E, whose cohomology class $c_1(E)$ is the first Chern class of E.

It is desirable to allow the hermitian structure to have singularity on a divisor. If $\phi_U = 0$ is the local representation of a divisor, we suppose

$$h_U = |\phi_U|^{2s}\, h'_U \,, \qquad h'_U > 0 \,, \tag{174}$$

where s is an integer. This generalized structure will be called *semi-hermitian*. If M is one-dimensional, the singularities of h_U are isolated and, relative to a suitable local coordinate ζ, h_U will be of the form

(175) $$h_U = |\zeta|^{2s} h'_U , \qquad h'_U > 0 .$$

The integer s is called the order of the singularity.

An application of Stokes' Theorem to (173) gives the theorem [17]:

Theorem 6.1. (Gauss-Bonnet) *Let* $\pi: E \to M$ *be a semi-hermitian holomorphic line bundle over a one-dimensional complex manifold* M . *Let* D *be a compact domain of* M *with smooth boundary* ∂D *and* $s: D \to E$ *be a holomorphic section such that*
1. ∂D *contains no singularity of the hermitian structure;*
2. $s(\partial D)$ *does not meet the zero section of the bundle. Then*

(176)
$$n(s) - n(h) = - \int_D \Omega + \frac{1}{2\pi} \int_{\partial D} d^c \log\|s\| , \qquad d^c = i(\bar{\partial}-\partial) ,$$

where $n(s)$ *is the number of zeroes of the section in* D, $n(h)$ *is the number of singularities of the semi-hermitian structure in* D, *and* $\|s\|$ *is the hermitian norm of the section on the boundary.*

We consider the complex projective space $P_n(C)$ of dimension n . To define it we take the complex vector space C_{n+1} of dimension $n + 1$ and identify its non-zero vectors which differ from each other by a factor. The identification

(177) $$\pi: C_{n+1} - \{0\} \to P_n(C)$$

defines a holomorphic line bundle over $P_n(C)$. To $x \in P_n(C)$, $\pi^{-1}(x)$ is a nonzero vector $Z = (z_0,\ldots,z_n)$ of C_{n+1} , determined up to a factor; Z will be called a homogeneous coordinate vector of x .

The geometry in $P_n(C)$ arises from the hermitian scalar product

(178) $$(Z,W) = z_0\bar{w}_0 + \ldots + z_n\bar{w}_n , \qquad W = (w_0,\ldots,w_n)$$

in C_{n+1} . We set

(179) $$|Z,W| = |(Z,W)| , \qquad |Z|^2 = (Z,Z) .$$

Then $|Z|$ defines an hermitian norm in the bundle (177). By (173) the Chern class of its dual bundle, the hyperplane section bundle, is represented by the curvature form

(180) $$\Omega = \frac{i}{\pi} \partial\bar{\partial} \log|Z| .$$

This form has the further property that it is positive definite, in the following sense: The complex structure on $P_n(C)$ sets up a one-one correspondence between real forms of bidegree (1,1) and the hermitian differential forms; the hermitian form corresponding to Ω is positive definite. We can therefore use it to define an hermitian structure on $P_n(C)$, which gives the classical Fubini-Study metric (Cf. [16]).

It can be verified that

(181) $$\int_{P_1} \Omega = 1 ,$$

so that $\{\Omega\}$, the cohomology class represented by Ω , is a generator of $H^2(P_n(C),Z)$. It follows that Ω^n is a volume element of $P_n(C)$, with total volume equal to 1.

Consider an algebraic curve

(182) $$f: M \to P_n(C) ,$$

where M is a compact one-dimensional complex manifold without boundary and f is holomorphic. The above discussion identifies the area of the curve with its order and we have the formula of Wirtinger:

(183) $$A(M) = \int_{f(M)} \Omega = n(f(M) \cap \alpha) = v(M) .$$

Here $A(M)$ is the area, $v(M)$ is the order of the curve, and $n(f(M) \cap \alpha)$ is the number of common points of $f(M)$ with any hyperplane α ; the equalities at the two ends of (183) are definitions.

The Gauss-Bonnet Theorem 6.1 extends this relationship to a compact domain D with boundary and we have the theorem:

Theorem 6.2 (Unintegrated first main theorem). *Let* M *be a one-dimensional complex manifold and* $f: M \to P_n(C)$ *be a holomorphic mapping. Let* D *be a compact domain of* M *with smooth boundary* ∂D *, such that* $f(\partial D)$ *does not meet a hyperplane* α . Then

(184) $$n(D,\alpha) - A(D) = \frac{1}{2\pi} \int_{f(\partial D)} d^C \log \frac{|Z,\alpha^{\perp}|}{|Z|\cdot|\alpha^{\perp}|} ,$$

where $n(D,\alpha)$ *is the number of points in* $f(D) \cap \alpha$ and $\alpha^{\perp}$ *is the "pole" of* α *, i.e., the point orthogonal to all points of* α .

Thus, while the formula (183) is not valid for a domain D with boundary, Theorem 6.2 gives a useful expression for the difference $n(D,\alpha)-A(D)$. When M is non-compact and D exhausts M , each of $n(D,\alpha)$ and $A(D)$ could become infinite and our main concern is to estimate their relative growth and the growth of other geometrical quantities which eventually enter into play. The quantity against which the growths are measured is the *exhaustion function*. It is by definition a smooth function $\tau: M \to R^+$ satisfying the conditions: (1) The mapping τ is proper, i.e., the inverse of a compact set is compact; (2) The critical points are isolated. An example of a function satisfying (2) is a real harmonic function.

Suppose τ be a harmonic exhaustion function of M . Let

(185) $$D_u = \{\zeta \in M | \tau(\zeta) \leqq u\} .$$

For simplicity we write

(186) $$n(D_u,\alpha) = n(u,\alpha) , \qquad A(D_u) = v_0(u) .$$

Then (184) can be written

(187) $$n(u,\alpha) - v_0(u) = \frac{1}{2\pi} \int_{f(\partial D_u)} d^C \log \frac{|Z(\zeta),\alpha^\perp|}{|Z(\zeta)|\cdot|\alpha^\perp|}, \qquad \zeta \in \partial D_u .$$

By a standard argument the integration and the differential operator d^C can be interchanged and we can integrate the above formula with respect to u. We put

(188) $$N(u,\alpha) = \int_0^u n(t,\alpha)dt, \qquad T_0(u) = \int_0^u v_0(t)dt$$

and

(189) $$m(u,\alpha) = \frac{1}{2\pi} \int_{\partial D_u} \log \frac{|Z(\zeta)|\cdot|\alpha^\perp|}{|Z(\zeta),\alpha^\perp|} d^C\tau \geqq 0, \qquad \zeta \in \partial D_u$$

Then the integration of (187) gives

(190) $$N(u,\alpha) + m(u,\alpha) = T_0(u) + m(0,\alpha) .$$

This is the *integrated form of the first main theorem.* As a corollary we have the *fundamental inequality*

(191) $$N(u,\alpha) < T_0(u) + \text{const.}$$

The function $T_0(u)$ is called the *order function.* Equation (191) shows that it dominates $N(u,\alpha)$ for all α.

The space of the hyperplanes of $P_n(C)$ has a measure defined to be the measure of their poles. That is, if η is a hyperplane, we define

(192) $$d\eta = d\eta^\perp, \qquad \eta^\perp = \text{pole of } \eta .$$

It is easy to prove:

Theorem 6.3 (Crofton-type formula). *Let* $f: M \to P_n(C)$ *be a compact holomorphic curve with or without boundary. Then*

$$\int n(f(M) \cap \eta)d\eta = A(M) , \tag{193}$$

where $A(M)$ *is the area of the curve.*

When it is applied to the inequality (191), we have:

Theorem 6.4 (Equidistribution in measure of holomorphic curves). *Let* $f: M \to P_n(C)$ *be a holomorphic curve, which has an exhaustion function* $u \to \infty$. *Then the set of hyperplanes* η *such that* $\eta \cap f(M) = \emptyset$ *is of measure zero.*

The strengthening of this theorem includes some of the most beautiful results in complex function theory. Following R. Nevanlinna we define the *defect* of hyperplane α by

$$\delta(\alpha) = \liminf_{u\to\infty} \frac{m(u,\alpha)}{T_0(u)} = 1 - \limsup_{u\to\infty} \frac{N(u,\alpha)}{T_0(u)} . \tag{194}$$

Then we have by (190)

$$0 \leqq \delta(\alpha) \leqq 1 \tag{195}$$

and $\delta(\alpha) = 1$ if $f(M) \cap \alpha = \emptyset$.

The fundamental theorem on value distributions can be stated as follows:

Theorem 6.5. *Let* $f: C \to P_n(C)$ *be a holomorphic curve which is non-degenerate (i.e., it does not lie in a linear space of lower dimension). Let* α_j , $1 \leqq j \leqq q$, *be* q *hyperplanes in general position. Then*

$$\sum_{1\leqq j\leqq q} \delta(\alpha_j) \leqq n + 1 . \tag{196}$$

The theorem was proved by R. Nevanlinna for the classical case $n = 1$ and by Ahlfors for general n. The proof in the general case is very long and we refer the reader to [32], [37] and, for the case $n = 2$, to [17]. Although the problem originates in analysis, it is most natural to regard it as a chapter in the complex differential geometry of curves.

There are many technical details in the proof. But two main ideas stand out as guideposts. The first idea is the consideration of the osculating spaces of all dimensions of the curve. By taking the osculating spaces of dimension k, we get a holomorphic curve f_k in the Grassmann manifold of all k-dimensional linear projective spaces in $P_n(C)$; $f_k(C)$ is called the kth associated curve. As in the case $k = 0$, this introduces the kth order function $T_k(u)$.

The second idea can be described as finding a lower bound for $N(u,\alpha)$, whereas the inequality (191) gives an upper bound. Following F. Nevanlinna and Ahlfors this is achieved by applying integral geometry with a singular density. The inequality in question can be written

(197)

$$(1-\lambda)\int_0^u dt \int_{D_t} \frac{|Z\wedge Z',\ Z\wedge\alpha^\perp|^2}{|Z|^4|Z\wedge\alpha^\perp|^2}\left(\frac{|Z|}{|Z,\alpha^\perp|}\right)^{2\lambda} |d\zeta d\bar{\zeta}| < BT_0(u) + B' ,$$

$$0 < \lambda < 1 , \qquad \zeta \in C$$

where B , B' are positive constants. This inequality, and its analogues for the associated curves, play a fundamental rôle in the proof of (196).

The broad outlines given above could well be the beginning of a long chapter on the global theory of holomorphic mappings of non-compact complex manifolds. We restrict ourselves in referring to the account of W. Stoll ([30]) and to recent studies by P. Griffiths and his coworkers ([36], [37]). It is conceivable that characteristic classes, in the spirit of this paper, will furnish the key to a satisfactory theory.

7. Chern-Simons Invariant of Three-dimensional Manifolds

In (125), §4 we defined an invariant $J = I(s)$ mod 1, for a compact oriented three-dimensional manifold. This has been called a Chern-Simons invariant. It has played an important role in both mathematics and physics. In fact, up to an additive constant it is the eta invariant of the manifold, as introduced by Atiyah, Patodi, and Singer via spectral theory [1]; cf. Ref to §7. It has also been used by W. Thurston in his theory of hyperbolic manifolds, while Robert Meyerhoff has shown that, for certain hyperbolic manifolds, it takes values which are dense on the unit circle. In mathematical physics the concept is found useful in quantum field theory, statistical mechanics, and the theory of anyons.

We begin by repeating its definition. Let M denote the manifold, oriented, and let P be the bundle of its orthonormal frames, so that we have

(198) $$\pi : P \longrightarrow M,$$

where π is the projection, mapping a frame $xe_1e_2e_3$ to its origin x. A section $s : M \longrightarrow P$ of the bundle satisfies the condition $\pi \circ s$ = identity and can be viewed as a field of frames. It is well known that in our case such a section always exists; we say that M is *parallelizable*.

To such a frame field the *Levi-Civita connection* of the metric is given by an antisymmetric matrix of one-forms:

(199) $$\varphi = (\varphi_{ij}), \qquad 1 \leq i,\ j \leq 3,$$

and its *curvature* by an antisymmetric matrix of two-forms:

(200) $$\Phi = (\Phi_{ij}), \qquad 1 \leq i,\ j \leq 3.$$

Throughout this section our small Latin indices will run from 1 to 3. We have

(201) $$d\varphi_{ik} = \sum_j \varphi_{ij} \wedge \varphi_{jk} + \Phi_{ik}.$$

We introduce the three-form

(202) $$T = \frac{1}{8\pi^2} \sum_{i,j,k} \left(\varphi_{ij} \wedge \Phi_{ij} - \tfrac{1}{3}\varphi_{ij} \wedge \varphi_{jk} \wedge \varphi_{ki}\right),$$

and consider the integral

(203) $$\Phi(s) = \int_M \tfrac{1}{2}T.$$

It will be proved below that for another section $t : M \to P$, $\Phi(t) - \Phi(s)$ is an integer, so that $\Phi(s)$ mod 1 is independent of s. This defines an invariant $J(M) \in \mathbb{R}/\mathbb{Z}$, where $J(M) = \Phi(s)$ mod 1.

In [3] we proved the theorems:

Theorem 7.1. *J(M) is a conformal invariant, i.e. it remains unchanged under a conformal transformation of the metric.*

Theorem 7.2. *J(M) has a critical value at M if and only if M is locally conformally flat.*

We wish to give direct proofs of these theorems in this section.

A) Family of Connections

We shall use arbitrary frame fields to develop the Riemannian geometry on M. Let $xe_1e_2e_3$ be a frame, and ω^1, ω^2, ω^3, its dual coframe. Let the inner product be

(204) $$\langle e_i, e_j\rangle = g_{ij}.$$

We introduce g^{ij} through the equations

(205) $$\sum g_{ij}g^{jk} = \delta_i^k$$

and use the g's to raise or lower indices, as in classical tensor analysis. Then the *connection forms* ω_i^j or ω_{ij} are determined, uniquely, by the equations

$$(206)\qquad d\omega^i = \sum \omega^j \wedge \omega_j^i, \qquad \omega_{ij} + \omega_{ji} = dg_{ij}.$$

The *curvature forms* are defined by

$$(207)\qquad \Omega_i^j = d\omega_i^j - \sum \omega_i^k \wedge \omega_k^j.$$

By exterior differentiation of (207) we get the *Bianchi identity*

$$(208)\qquad d\Omega_i^j = \sum \omega_i^k \wedge \Omega_k^j - \sum \Omega_i^k \wedge \omega_k^j.$$

To avoid confusion notice our convention that the upper index in ω_i^j, Ω_i^j is the second index. Thus $\sum_j \omega_i^j g_{jk} = \omega_{ik}$ ($\neq \omega_{ki}$ in general).

We introduce the cubic form

$$(209)\quad 8\pi^2 T = -\frac{1}{3}\sum \omega_i^j \wedge \omega_j^k \wedge \omega_k^i + \sum \omega_i^i \wedge \Omega_j^j - \sum \omega_i^j \wedge \Omega_j^i.$$

On our three-manifold M, T is of course closed. But the basic reason for its importance and interesting properties is that formally by (207), (208), we have

$$(210)\qquad \begin{aligned} 8\pi^2 dT &= \sum \left(\Omega_i^i \wedge \Omega_j^j - \Omega_i^j \wedge \Omega_j^i\right) \\ &= \sum \delta_{i_1 i_2}^{j_1 j_2} \Omega_{j_1}^{i_1} \wedge \Omega_{j_2}^{i_2}, \end{aligned}$$

which is the *first Pontrjagin form*.

Consider a family of connections on M, depending on a parameter τ. Then ω_i^j, Ω_i^j, T all involve τ, and we have the fundamental formula

$$(211)\qquad \begin{aligned} 8\pi^2 \frac{\partial T}{\partial \tau} &= -d\left\{\sum \omega_i^i \wedge \frac{\partial \omega_j^j}{\partial \tau} - \sum \omega_i^j \frac{\partial \omega_j^i}{\partial \tau}\right\} \\ &\quad + 2\sum\left(\frac{\partial \omega_i^i}{\partial \tau} \wedge \Omega_j^j - \frac{\partial \omega_i^j}{\partial \tau} \wedge \Omega_j^i\right). \end{aligned}$$

The proof of (211) is straightforward. It follows by differentiation of (209), and using the formulas obtained by differentiation of (206), (207) with respect to τ. It is useful to observe that the last term is a polarization of the Pontrjagin form.

Let P' be the bundle of all frames of M, so that we have

(212)
$$\begin{array}{ccc} P & \xrightarrow{i} & P' \\ & {\scriptstyle\pi}\searrow \swarrow{\scriptstyle\pi'} & \\ & M & \end{array},$$

where i is the inclusion. Then T in (209) can be considered as a form in P' and its pull-back $i*T$ is the T given by (202). We will make such identifications when there is no danger of confusion.

We consider $\Phi(s)$ defined by (203). When t is another section, then $t(M) - s(M)$, as a three-dimensional cycle is homologous, modulo torsion, to an integral multiple of the fiber P_x, $x \in M$. But P_x is topologically SO(3) and ω_{ij} reduces on it to its Maurer-Cartan forms. If $j : P_x = SO(3) \rightarrow P$ is the inclusion,

$$\pm\frac{1}{2}\ j * T = \frac{1}{8\pi^2}\omega_{12} \wedge \omega_{23} \wedge \omega_{31},$$

whose integral over P_x is 1. Hence $\Phi(s)$ mod 1 is independent of s.

We wish to clarify the relation between orthonormal frame fields and arbitrary frame fields. By the Schmidt orthogonalization process P is a retract of P', under which the origin of the frame is fixed. The retraction we denote by $r : P' \rightarrow P$. We consider the form T defined in (209) to be in P'. If $s' : M \rightarrow P'$ is a section, then $s = r \circ s'$ is also a section and they are homotopic through sections. Since $dT = 0$, we have

(213)
$$\int_{s'M} T = \int_{sM} i*T.$$

Hence J(M) can be computed through an arbitrary frame field by the left-hand side of the last equation.

B) Proofs of the Theorems

In view of the above remark we can, for local considerations, use a local coordinate system u^i and the resulting natural frame field $\partial/\partial u^i$. We shall summarize the well-known formulas, which are

$$\omega_i^j = \sum \Gamma_{ik}^j du^k,$$

$$\Gamma_{ik}^j = \sum g^{jl}\Gamma_{ilk}, \tag{214}$$

$$\Gamma_{ilk} = \frac{1}{2}\left(\frac{\partial g_{il}}{\partial u^k} + \frac{\partial g_{kl}}{\partial u^i} - \frac{\partial g_{ik}}{\partial u^l}\right),$$

$$\Omega_i^j = \frac{1}{2}\sum R_{ikl}^j du^k \wedge du^l.$$

The R_{ijkl} satisfy the symmetry relations

$$R_{ijkl} = -R_{jikl} = -R_{ijlk}, \quad R_{ijkl} = R_{klij}, \tag{215}$$

$$R_{ijkl} + R_{iklj} + R_{iljk} = 0, \quad R_{ijkl,h} + R_{ijlh,k} + R_{ijhk,l} = 0,$$

where the comma denotes covariant differentiation. They imply

$$\Omega_{ij} + \Omega_{ji} = 0, \quad \sum \Omega_{ij} \wedge dx^j = 0, \tag{216}$$

and

$$\sum \Omega_i^i = \sum g^{ik}\Omega_{ik} = 0. \tag{217}$$

We also introduce the *Ricci curvature* and the *scalar curvature by*

$$R_k^i = \sum_j R_{kj}^{ij}, \quad R = \sum R_i^i. \tag{218}$$

For treatment of the conformal geometry we define

$$C_{kl}^j = R_{k,l}^j - R_{l,k}^j - \frac{1}{4}(\delta_k^j R_{,l} - \delta_l^j R_{,k}),$$

$$C_{jkl} = R_{jk,l} - R_{jl,k} - \frac{1}{4}(g_{jk}R_{,l} - g_{jl}R_{,k}). \tag{219}$$

Then the Bianchi identities give

$$\sum C_{kj}^j = 0, \quad C_{jkl} + C_{klj} + C_{ljk} = 0. \tag{220}$$

These relations imply that the matrix

(221)
$$C = \begin{bmatrix} C^1_{23} & C^1_{31} & C^1_{12} \\ C^2_{23} & C^2_{31} & C^2_{12} \\ C^3_{23} & C^3_{31} & C^3_{12} \end{bmatrix}$$

is symmetric and that the matrix GC, where $G = (g_{ij})$, has trace zero.

Schouten proved [4, p. 92] that the three-dimensional manifold M is conformally flat, if and only if $C = 0$, i.e. $C_{ijk} = 0$.

By integrating (211), we get

(222)
$$8\pi^2 \frac{\partial}{\partial\tau}\int_M T = 2\int_M \Delta,$$

where

(223)
$$\Delta = \sum\left(\frac{\partial\omega^i_i}{\partial\tau}\wedge\Omega^j_j - \frac{\partial\omega^j_i}{\partial\tau}\wedge\Omega^i_j\right) = -\sum\frac{\partial\omega^j_i}{\partial\tau}\wedge\Omega^i_j,$$

by (217).

We consider a family of metrics $g_{ij}(\tau)$ and put

(224)
$$v_{ij} = \frac{\partial g_{ij}}{\partial\tau}.$$

To prove Theorem 7.1 we suppose this is a conformal family of metrics, i.e.

(225)
$$v_{ij} = \sigma g_{ij}.$$

From (214) we find

(226)
$$\frac{\partial\omega^j_i}{\partial\tau} = \frac{1}{2}\sum(\delta^j_i\sigma_k + \delta^j_k\sigma_i - g_{ik}g^{jl}\sigma_l)dx^k,$$

where $\sigma_k = \partial\sigma/\partial x^k$. By the second equation of (216) and the equation (217) we find $\Delta = 0$. This proves that $\int_M T$ is independent of τ, and hence Theorem 7.1.

To prove Theorem 7.2 we consider v_{ij} such that the trace

$\sum v_i^i = 0$. Geometrically this means that we consider the tangent space of the space of conformal structures on M. From (214) we find

(227)

$$\frac{\partial \omega_i^j}{\partial \tau} = \sum g^{jl}\left\{-\sum v_{lk}\omega_i^k + \frac{1}{2}\sum\left(\frac{\partial v_{il}}{\partial u^k} + \frac{\partial v_{kl}}{\partial u^i} - \frac{\partial v_{ik}}{\partial u^l}\right)du^k\right\}.$$

It follows that

(228)

$$\Delta = \sum_{i,j}\Omega^{ij}\left\{-\sum v_{jk}\omega_i^k + \frac{1}{2}dv_{ij} + \frac{1}{2}\sum_k\left(\frac{\partial v_{kj}}{\partial u^i} - \frac{\partial v_{ki}}{\partial u^j}\right)du^k\right\}.$$

The term in the middle is zero, because Ω^{ij} is antisymmetric and dv_{ij} is symmetric in i, j. To the integral of the last term we apply Stokes theorem to reduce it to an integral involving only the v_{ij}, and not their derivatives. We will omit the details of this computation. The result is that the condition

$$\frac{\partial}{\partial \tau}\int_M T = 0$$

is equivalent to

(229) $$\int_M \mathrm{Tr}(VC)\ du^1 \wedge du^2 \wedge du^3 = 0.$$

If the metric is conformally flat, we have C = 0 and hence the vanishing of the above integral.

Conversely, at a critical point of Φ we must have Tr(VC) = 0 for all symmetric V satisfying $\mathrm{Tr}(VG^{-1}) = 0$. Hence C is a multiple of G^{-1} or GC is a multiple of the unit matrix. But GC has trace zero. Hence it must itself be zero and we have C = 0. This proves Theorem 7.2.

References

1. L. Ahlfors, *The theory of meromorphic curves,* Acta Soc. Sci. Fenn. Nova Ser. A 3 (1941), no. 4, 1-31.

2. M. F. Atiyah and I. M. Singer, *The Index of elliptic operators: III,* Ann. of Math. 87 (1968), 546-604.

3. A. Avez, *Characteristic classes and Weyl tensor: applications to general relativity,* Proc. Nat. Acad. Sci. (USA) 66 (1970), 265-268.

4. P. F. Baum, *Vector fields and Gauss-Bonnet,* Bull. Amer. Math. Soc. 76 (1970), 1202-1211.

5. P. F. Baum and R. Bott, *On the zeroes of meromorphic vector fields, Essays on topology and related topics dedicated to G. de Rham,* 1970, 29-47.

6. P. F. Baum and J. Cheeger, *Infinitesimal isometries and Pontrjagin numbers,* Topology 8 (1969), 173-193.

7. A. Borel and F. Hirzebruch, *Characteristic classes and homogeneous spaces,* Amer. J. Math. 80 (1958), 458-538; 81 (1959), 315-382; 82 (1960), 491-504.

8. R. Bott, *Vector fields and characteristic numbers,* Michigan Math. J. 14 (1967), 231-244.

9. _______, *A residue formula for holomorphic vector fields,* J. Diff. Geom. 1 (1967), 311-330.

10. R. Bott, *On a topological obstruction to integrability,* Proc. Symp. in Pure Math., Amer. Math. Soc. 16 (1970), 127-131.

11. H Cartan, *Notion d'algèbre différentielle; applications aux groupes de Lie et aux variétés ou opère un groupe de Lie,* Coll. de Topologie, Bruxelles (1950), 15-27.

12. _______, *La transgression dans un groupe de Lie et dans un espace fibré principal,* ibid, 57-71.

13. S. Chern, *Characteristic classes of hermitian manifolds,* Annals of Math. 47 (1946), 85-121.

14. _______, *Topics in differential geometry,* Inst. for Adv. Study, Princeton, 1951.

15. S. Chern, *Differential geometry of fiber bundles.* Int. Cong. of Math., II (1950), 397-411.

16. _______, *Complex manifolds without potential theory,* van Nostrand 1967; also, this book.

17. _______, *Holomorphic curves in the plane,* Differential geometry in honor of Yano, Tokyo, 1972, 73-94.

18. _______, and J. Simons, *Some cohomology classes in principal fiber bundles and their application to riemannian geometry*, Proc. Nat. Acad. Sci. (USA) 68 (1971), 791-794.

19. _______, and J. Simons, *Characteristic forms and geometrical invariants*, Annals of Math 99 (1974), 48-69.

20. L. P. Eisenhart, *Riemannian geometry*, Princeton, 1949.

21. F. Hirzebruch, *Topological methods in algebraic geometry*, Springer, 1966.

22. H. Hopf, *Vektorfelder in Mannigfaltigkeiten*, Math. Annalen 96 (1927), 225-250.

23. D. Husemoller, *Fibre bundles*, McGraw-Hill, 1966.

24. S. Kobayashi and K. Nomizu, *Foundations of differential geometry*, vol. 2, Interscience, 1969.

25. S. Kobayashi and T. Ochiai, *G-structures of order two and transgression operators*, J. Diff. Geom. 6 (1971), 213-230.

26. J. Milnor and J. D. Stasheff, *Characteristic classes*, Annals of Math. Studies 76, Princeton, 1974.

27. L. Pontrjagin, *Some topological invariants of closed riemannian manifolds*, Izvestiya Akad. Nauk SSSR Ser. Mat. 13 (1949), 125-162.

28. N. Steenrod, *The topology of fibre bundles*, Princeton 1951.

29. E. Stiefel, *Richtungsfelder und Fernparallelismus in Mannigfaltigkeiten*, Comm. Math. Helv. 8 (1936), 3-51.

30. W. Stoll, *Value distribution of holomorphic maps into compact complex manifolds*, Springer notes vol. 135 (1970).

31. H. Whitney, *On the topology of differntiable manifolds*, Lectures in Topology, Univ. of Michigan Press, 1941.

32. Hung-Hsi Wu, *The equidistribution theory of holomorphic curves*, Annals of Math. Studies 64, Princeton, 1970.

33. Wen-Tsun Wu, *Sur les espaces fibrés*, Act. Sci. et Indus. 1183 (1952).

34. S. Chern, *Meromorphic vector fields and characteristic numbers*, Scripta Math. 29 (1973), 243-251.

35. R. Bott and A. Haefliger, *On characteristic classes of Γ-foliations*, Bull. AMS 78 (1972), 1039-1044.

36. J. Carlson and P. Griffiths, *A defect relation for equidimensional holomorphic mappings between algebraic varieties*, Annals of Math. 95 (1972), 557-584.

37. M. Cowen and P. Griffiths, *Holomorphic curves and metrics of negative curvature*, J d'Analyse Math 29 (1976), 93-153.

References to §7

1. M. F. Atiyah, V. K. Patodi and I. M. Singer, ''Spectral asymmetry and Riemannian geometry, II,'' *Math. Proc. Cambridge Philos. Soc.* 78 (1975), 405-432.

2. S. Axelrod and I. M. Singer, ''Chern-Simons perturbation theory, II,'' *Jour. of Differential Geometry* 39 (1994), 173-213.

3. S. S. Chern and J. Simons, ''Characteristic forms and geometrical invariants,'' *Ann. of Math.* 99 (1974), 48-69 or S. S. Chern, *Selected Papers* (Springer, Berlin, 1978), 444-465.

4. L. P. Eisenhart, *Riemannian Geometry* (Princeton University Press, Princeton, 1949).

5. J. Frölich, ''The fractional quantum Hall effect, Chern-Simons theory, and integral lattices,'' *Intrn'tl. Congress of Math.*, Zurich 1994.

6. R. Jackiw, ''Solitons in Chern-Simons/Anyon systems,'' in *Physics and Math of Anyons*, edited by Chern-Chu-Ting (World Scientific, 1991), 127-137.

7. R. Meyerhoff, ''The Chern-Simons invariant for hyperbolic 3-manifolds,'' thesis (Princeton University, Princeton, 1981).

8. W. Thurston, ''Three-dimensional manifolds, Kleinian groups, and hyperbolic geometry,'' in *The Mathematical Heritage of Henri Poincaré*, (American Mathematical Society, Providence, RI, 1983), 87-111.

9. E. Witten, ''Quantum field theory and the Jones polynomial,'' in *Braid Groups, Knot Theory, and Statistical Mechanics*, edited by C. N. Yang and M. L. Ge, (World Scientific, 1989), 239-307.

Index

9 780387 904221